T0256933

**ANNALS OF MATHEMATICS STUDIES**
**NUMBER 6**

ANNALS OF MATHEMATICS STUDIES

*Edited by Emil Artin and Marston Morse*

1. Algebraic Theory of Numbers, *by* HERMANN WEYL
3. Consistency of the Continuum Hypothesis, *by* KURT GÖDEL
6. The Calculi of Lambda-Conversion, *by* ALONZO CHURCH
7 Finite Dimensional Vector Spaces, *by* PAUL R. HALMOS
10. Topics in Topology, *by* SOLOMON LEFSCHETZ
11. Introduction to Nonlinear Mechanics, *by* N. KRYLOFF *and* N. BOGOLIUBOFF
14. Lectures on Differential Equations, *by* SOLOMON LEFSCHETZ
15. Topological Methods in the Theory of Functions of a Complex Variable, *by* MARSTON MORSE
16. Transcendental Numbers, *by* CARL LUDWIG SIEGEL
17. Problème Général de la Stabilité du Mouvement, *by* M. A. LIAPOUNOFF
18. A Unified Theory of Special Functions, *by* C. A. TRUESDELL
19. Fourier Transforms, *by* S. BOCHNER *and* K. CHANDRASEKHARAN
20. Contributions to the Theory of Nonlinear Oscillations, *edited by* S. LEFSCHETZ
21. Functional Operators, Vol. I, *by* JOHN VON NEUMANN
22. Functional Operators, Vol. II, *by* JOHN VON NEUMANN
23. Existence Theorems in Partial Differential Equations, *by* DOROTHY L. BERNSTEIN
24. Contributions to the Theory of Games, *edited by* H. W. KUHN *and* A. W. TUCKER
25. Contributions to Fourier Analysis, *by* A. ZYGMUND, W. TRANSUE, M. MORSE, A. P. CALDERON, *and* S. BOCHNER
26. A Theory of Cross-Spaces, *by* ROBERT SCHATTEN
27. Isoperimetric Inequalities in Mathematical Physics, *by* G. POLYA *and* G. SZEGO

# THE CALCULI OF LAMBDA-CONVERSION

BY

ALONZO CHURCH

PRINCETON
PRINCETON UNIVERSITY PRESS
LONDON: HUMPHREY MILFORD
OXFORD UNIVERSITY PRESS
1941

Second Printing 1951

Lithoprinted in U.S.A.
EDWARDS BROTHERS, INC.
ANN ARBOR, MICHIGAN
1951

CONTENTS

Chapter I

# INTRODUCTORY

1. THE CONCEPT OF A FUNCTION. Underlying the formal calculi which we shall develop is the concept of a function, as it appears in various branches of mathematics, either under that name or under one of the synonymous names, "operation" or "transformation." The study of the general properties of functions, independently of their appearance in any particular mathematical (or other) domain, belongs to formal logic or lies on the boundary line between logic and mathematics. This study is the original motivation for the calculi — but they are so formulated that it is possible to abstract from the intended meaning and regard them merely as formal systems.

A <u>function</u> is a rule of correspondence by which when anything is given (as <u>argument</u>) another thing (the <u>value</u> of the function for that argument) may be obtained. That is, a function is an operation which may be applied on one thing (the argument) to yield another thing (the value of the function). It is not, however, required that the operation shall necessarily be applicable to everything whatsoever; but for each function there is a class, or range, of possible arguments -- the class of things to which the operation is significantly applicable -- and this we shall call the <u>range of arguments</u>, or <u>range of the independent variable</u>, for that function. The class of all values of the function, obtained by taking all possible arguments, will be called the <u>range of values</u>, or <u>range of the dependent variable</u>.

If $f$ denotes a particular function, we shall use the notation $(fa)$ for the value of the function $f$ for the argument $a$. If $a$ does not belong to the range of arguments of $f$, the notation $(fa)$ shall be meaningless.

It is, of course, not excluded that the range of arguments or range of values of a function should consist wholly or partly of functions. The derivative, as this notion appears in the elementary differential calculus, is a familiar mathematical example of a function for which both ranges consist of functions. Or, turning to the integral calculus, if in the expression $\int_0^1 (fx)dx$ we take the function $f$ as independent variable, we are led to a function for which the range of arguments consists of functions and the range of values, of numbers. Formal logic provides other examples; thus the existential quantifier, according to the present account, is a function for which the range of arguments consists of propositional functions, and the range of values consists of truth-values.

In particular it is not excluded that one of the elements of the range of arguments of a function $f$ should be the function $f$ itself. This possibility has frequently been denied, and indeed, if a function is defined as a correspondence between two previously given ranges, the reason for the denial is clear. Here, however, we regard the operation or rule of correspondence, which constitutes the function, as being first given, and the range of arguments then determined as consisting of the things to which the operation is applicable. This is a departure from the point of view usual in mathematics, but it is a departure which is natural in passing from consideration of functions in a special domain to the consideration of function in general, and it finds support in consistency theorems which will be proved below.

The identity function $I$ is defined by the rule that $(Ix)$ is $x$, whatever $x$ may be; then in particular $(II)$ is $I$. If a function $H$ is defined by the rule that $(Hx)$ is $I$, whatever $x$ may be, then in particular $(HH)$ is $I$. If $\Sigma$ is the existential quantifier, then $(\Sigma\Sigma)$ is the truth-value truth.

The functions $I$ and $H$ may also be cited as examples of functions for which the range of arguments consists of all things whatsoever.

2. EXTENSION AND INTENSION. The foregoing discussion leaves it undetermined under what circumstances two functions shall be considered the same.

The most immediate and, from some points of view, the best way to settle this question is to specify that two functions $f$ and $g$ are the same if they have the same range of arguments and, for every element $a$ that belongs to this range, $(fa)$ is the same as $(ga)$. When this is done we shall say that we are dealing with functions in extension.

It is possible, however, to allow two functions to be different on the ground that the rule of correspondence is different in meaning in the two cases although always yielding the same result when applied to any particular argument. When this is done we shall say that we are dealing with functions in intension. The notion of difference in meaning between two rules of correspondence is a vague one, but, in terms of some system of notation, it can be made exact in various ways. We shall not attempt to decide what is the true notion of difference in meaning but shall speak of functions in intension in any case where a more severe criterion of identity is adopted than for functions in extension. There is thus not one notion of function in intension, but many notions, involving various degrees of intensionality.

In the calculus of $\lambda$-conversion and the calculus of restricted $\lambda$-$K$-conversion, as developed below, it is possible, if desired, to interpret the expressions of the calculus as denoting functions in extension. However, in the calculus of $\lambda$-$\delta$-conversion, where the notion of identity of functions is introduced into the system by the symbol $\delta$, it is necessary, in order to preserve the finitary character of the transformation rules, so to formulate these rules that an interpretation by functions in extension becomes impossible. The expressions which appear in the calculus of $\lambda$-$\delta$-conversion are interpretable as denoting functions in intension of an appropriate kind.

3. FUNCTIONS OF SEVERAL VARIABLES. So far we have tacitly restricted the term "function" to functions of one variable (or, of one argument). It is desirable, however, for each positive integer $n$, to have the notion of a function of $n$ variables. And, in order to avoid the introduction of a separate primitive idea for each $n$, it is desirable to find a means of explaining functions of $n$ variables as particular cases of functions of one variable. For our present purpose, the most convenient and natural method of doing this is to adopt an idea of Schönfinkel [49], according to which a function of two variables is regarded as a function of one variable whose values are functions of one

variable, a function of three variables as a function of one variable whose values are functions of two variables, and so on.

Thus if $f$ denotes a particular function of two variables, the notation $((fa)b)$ -- which we shall frequently abbreviate as $(fab)$ or $fab$ -- represents the value of $f$ for the arguments $a, b$. The notation $(fa)$ -- which we shall frequently abbreviate as $fa$ -- represents a function of one variable, whose value for any argument $x$ is $fax$. The function $f$ has a range of arguments, and the notation $fa$ is meaningful only when $a$ belongs to that range; the function $fa$ again has a range of arguments, which is, in general, different for different elements $a$, and the notation $fab$ is meaningful only when $b$ belongs to the range of arguments of $fa$.

Similarly, if $f$ denotes a function of three variables, $(((fa)b)c)$ or $fabc$ denotes the value of $f$ for the arguments $a, b, c$, $fa$ denoting a certain function of two variables, and $((fa)b)$ or $fab$ denoting a certain function of one variable -- and so on.

(According to another scheme, which is the better one for certain purposes, a function of two variables is regarded as a function (of one variable) whose arguments are ordered pairs, a function of three variables as a function whose arguments are ordered triads, and so on. This other concept of a function of several variables is not however, excluded here. For, as will appear below, the notions of ordered pair, ordered triad, etc., are definable by means of abstraction (§4) and the Schönfinkel concept of a function of severable variables; and thus functions of several variables in the other sense are also provided for.)

An example of a function of two variables (in the sense of Schönfinkel) is the <u>constancy function $K$</u>, defined by the rule that $Kxy$ is $x$, whatever $x$ and $y$ may be. We have, for instance that $KII$ is $I$, $KHI$ is $H$, and so on. Also $KI$ is $H$ (where $H$ is the function defined above in §1). Similarly $KK$ is a function whose value is constant and equal to $K$.

Another example of a function of two variables is the function whose value for the arguments $f$, $x$ is $(fx)$; for reasons which will appear later we designate this function by the symbol $1$. The function $1$, regarded as a function of one variable, is a kind of identity function, since the notation $(1f)$

whenever significant, denotes the same function as $f$; the functions $I$ and $1$ are not, however, the same function, since the range of arguments consists in one case of all things whatever, in the other case merely of all functions.

Other examples of functions of two or more variables are the function $H$, already defined, and the functions $T$, $J$, $B$, $C$, $W$, $S$, defined respectively by the rules that $Txf$ is $(fx)$, $Jfxyz$ is $fx(fzy)$, $Bfgx$ is $f(gx)$, $Cfxy$ is $(fyx)$, $Wfx$ is $(fxx)$, $Snfx$ is $f(nfx)$.

Of these, $B$ and $C$ may be more familiar to the reader under other names, as the <u>product</u> or <u>resultant</u> of two transformations $f$ and $g$, and as the <u>converse</u> of a function of two variables $f$. To say that $BII$ is $I$ is to say that the product of the identity transformation by the identity transformation is the identity transformation, whatever the domain within which transformations are being considered; to say that $B11$ is $1$ is to say that within any domain consisting entirely of functions the product of the identity transformation by itself is the identity transformation. $BI$ is $1$, since it is the operation of composition with the identity transformation, and thus an identity operation, but one applicable only to transformations.

The reader may further verify that $CK$ is $H$, $CT$ is $1$, $C1$ is $T$, $CI$ is $T$ -- that $1$ and $I$ have the same converse is explained by the fact that, while not the same function, they have the same effect in all cases where they can significantly be applied to two arguments. The function $BCC$, the converse of the converse, has the effect of an identity when applied to a function of two variables, but when applied to a function of one variable it has the effect of so restricting the range of arguments as to transform the function into a function of two variables (if possible); thus $BCCI$ is $1$.

There are many similar relations between these functions, some of them quite complicated.

4. ABSTRACTION. For our present purpose it is necessary to distinguish carefully between a symbol or expression which denotes a function and an expression which contains a variable and denotes ambiguously some value of the function -- a distinction which is more or less obscured in the usual language of

mathematical function theory.

To take an example from the theory of functions of natural numbers, consider the expression $(x^2+x)^2$. If we say, "$(x^2+x)^2$ is greater than 1,000," we make a statement which depends on $x$ and actually has no meaning unless $x$ is determined as some particular natural number. On the other hand, if we say, "$(x^2+x)^2$ is a primitive recursive function," we make a definite statement whose meaning in no way depends on a determination of the variable $x$ (so that in this case $x$ plays the rôle of an apparent, or bound, variable). The difference between the two cases is that in the first case the expression $(x^2+x)^2$ serves as an ambiguous, or variable, denotation of a natural number, while in the second case it serves as the denotation of a particular function. We shall hereafter distinguish by using $(x^2+x)^2$ when we intend an ambiguous denotation of a natural number, but $(\lambda x(x^2+x)^2)$ as the denotation of the corresponding function -- and likewise in other cases.

(It is, of course, irrelevant here that the notation $(x^2+x)^2$ is commonly used also for a certain function of real numbers, a certain function of complex numbers, etc. In a logically exact notation the functions, addition of natural numbers, addition of real numbers, addition of complex numbers, would be denoted by different symbols, say $+_n$, $+_r$, $+_c$; and the three functions, square of a natural number, square of a real number, square of a complex number, would be similarly distinguished. The uncertainty as to the exact meaning of the notation $(x^2+x)^2$, and the consequent uncertainty as to the range of arguments of the function $(\lambda x(x^2+x)^2)$, would then disappear.)

In general, if $\boldsymbol{M}$ is an expression containing a variable $x$ (as a free variable, i.e., in such a way that the meaning of $\boldsymbol{M}$ depends on a determination of $x$), then $(\lambda x\boldsymbol{M})$ denotes a function whose value, for an argument $a$, is denoted by the result of substituting (a symbol denoting) $a$ for $x$ in $\boldsymbol{M}$. The range of arguments of the function $(\lambda x\boldsymbol{M})$ consists of all objects $a$ such that the expression $\boldsymbol{M}$ has a meaning when (a symbol denoting) $a$ is substituted for $x$.

If $\boldsymbol{M}$ does not contain the variable $x$ (as a free variable), then $(\lambda x\boldsymbol{M})$ might be used to denote a function whose value is constant and equal to (the thing denoted by) $\boldsymbol{M}$, and whose range of arguments consists of all things. This usage is contemplated below in connection with the calculi of $\lambda$-$K$-conversion, but is excluded from the calculi of $\lambda$-conversion and $\lambda$-$\delta$-conversion -- for technical reasons which will appear.

Notice that, although $x$ occur as a free variable in $M$, nevertheless, in the expression $(\lambda x M)$, $x$ is a bound, or apparent, variable. Example: the equation $(x^2+x)^2 = (y^2+y)^2$ expresses a relation between the natural numbers denoted by $x$ and $y$ and its truth depends on a determination of $x$ and of $y$ (in fact, it is true if and only if $x$ and $y$ are determined as denoting the same natural number); but the equation $(\lambda x(x^2+x)^2) = (\lambda y(y^2+y)^2)$ expresses a particular proposition -- namely that $(\lambda x(x^2+x)^2)$ is the same function as $(\lambda y(y^2+y)^2)$ -- and it is true (there is no question of a determination of $x$ and $y$).

Notice also that $\lambda$, or $\lambda x$, is not the name of any function or other abstract object, but is an incomplete symbol -- i.e., the symbol has no meaning alone, but appropriately formed expressions containing the symbol have a meaning. We call the symbol $\lambda x$ an abstraction operator, and speak of the function which is denoted by $(\lambda x M)$ as obtained from the expression $M$ by abstraction.

The expression $(\lambda x(\lambda y M))$, which we shall often abbreviate as $(\lambda xy.M)$, denotes a function whose value, for an argument denoted by $x$, is denoted by $(\lambda y M)$ -- thus a function whose values are functions, or a function of two variables. The expression $(\lambda y(\lambda x M))$, abbreviated as $(\lambda yx.M)$, denotes the converse function to that denoted by $(\lambda xy.M)$. Similarly $(\lambda x(\lambda y(\lambda z M)))$, abbreviated as $(\lambda xyz.M)$, denotes a function of three variables, and so on.

Functions introduced in previous sections as examples can now be expressed, if desired, by means of abstraction operators. For instance, $I$ is $(\lambda xx)$; $J$ is $(\lambda fxyz.fx(fzy))$; $S$ is $(\lambda nfx.f(nfx))$; $H$ is $(\lambda xI)$, or $(\lambda x(\lambda yy))$, or $(\lambda xy.y)$; $K$ is $(\lambda xy.x)$; $1$ is $(\lambda fx.fx)$.

# Chapter II

# LAMBDA-CONVERSION

5. PRIMITIVE SYMBOLS, AND FORMULAS. We turn now to the development of a formal system, which we shall call the calculus of λ-conversion, and which shall have as a possible interpretation or application the system of ideas about functions described in Chapter I.

The primitive symbols of this calculus are three symbols,

$$\lambda, \; (, \; ),$$

which we shall call improper symbols, and an infinite list of symbols,

$$a, b, c, \ldots, x, y, z, \bar{a}, \bar{b}, \ldots, \bar{z}, \bar{\bar{a}}, \ldots,$$

which we shall call variables. The order in which the variables appear in this originally given infinite list shall be called their alphabetical order.

A formula is any finite sequence of primitive symbols. Certain formulas are distinguished as well-formed formulas, and each occurrence of a variable in a well-formed formula is distinguished as free or bound, in accordance with the following rules (1-4), which constitute a definition of these terms by recursion:

1. A variable $\mathbf{x}$ is a well-formed formula, and the occurrence of the variable $\mathbf{x}$ in this formula is free.

2. If $\mathbf{F}$ and $\mathbf{A}$ are well-formed, $(\mathbf{FA})$ is well-formed, and an occurrence of a variable $\mathbf{y}$ in $\mathbf{F}$ is free or bound in $(\mathbf{FA})$ according as it is free or bound in $\mathbf{F}$, and an occurrence of a variable $\mathbf{y}$ in $\mathbf{A}$ is free or bound in $(\mathbf{FA})$ according as it is free or bound in $\mathbf{A}$.

3. If $\mathbf{M}$ is well-formed and contains at least one free occurrence of $\mathbf{x}$, then $(\lambda\mathbf{xM})$ is well-formed, and an occurrence

of a variable $\mathbf{y}$, other than $\mathbf{x}$, in $(\lambda\mathbf{x}\mathbf{M})$ is free or bound in $(\lambda\mathbf{x}\mathbf{M})$ according as it is free or bound in $\mathbf{M}$. All occurrences of $\mathbf{x}$ in $(\lambda\mathbf{x}\mathbf{M})$ are bound.

4. A formula is well-formed, and an occurrence of a variable in it is free, or is bound, only when this follows from 1-3.

The free variables of a formula are the variables which have at least one free occurrence in the formula. The bound variables of a formula are the variables which have at least one bound occurrence in the formula.

Hereafter (as was just done in the statement of the rules 1-4) we shall use bold capital letters to stand for variable or undetermined formulas, and bold small letters to stand for variable or undetermined variables. Unless otherwise indicated in a particular case, it is to be understood that formulas represented by bold capital letters are well-formed formulas. Bold letters are thus not part of the calculus which we are developing but are a device for use in talking about the calculus: they belong, not to the system itself, but to the metamathematics or syntax of the system.

Another syntactical notation which we shall use is the notation,

$$\mathrm{S}^{\mathbf{x}}_{\mathbf{N}}\mathbf{M}|$$

which shall stand for the formula which results by substitution of $\mathbf{N}$ for $\mathbf{x}$ throughout $\mathbf{M}$. This formula is well-formed, except in the case that $\mathbf{x}$ is a bound variable of $\mathbf{M}$ and $\mathbf{N}$ is other than a single variable -- see §7. (In the special case that $\mathbf{x}$ does not occur in $\mathbf{M}$, it is the same formula as $\mathbf{M}$.)

For brevity and perspicuity in dealing with particular well-formed formulas, we often do not write them in full but employ various abbreviations.

One method of abbreviation is by means of a nominal definition, which introduces a particular new symbol to replace or stand for a particular well-formed formula. We indicate such a nominal definition by an arrow, pointing from the new symbol which is being introduced to the well-formed formula which it is to replace (the arrow may be read "stands for"). As an example we make at once the nominal definition:

$$I \rightarrow (\lambda a a).$$

This means that $I$ will be used as an abbreviation for $(\lambda a a)$ -- and consequently that $(II)$ will be used as an abbreviation for $((\lambda a a)(\lambda a a))$, $(\lambda a(a I))$ as an abbreviation for $(\lambda a(a (\lambda a a)))$, etc.

Another method of abbreviation is by means of a schematic definition, which introduces a class of new expressions of a certain form, specifying a scheme according to which each of the new expressions stands for a corresponding well-formed formula. Such a schematic definition is indicated in a similar fashion by an arrow, but the expressions on each side of the arrow contain bold letters. When a bold small letter -- one or several -- occurs in the expression following the arrow (the definiens) but not in the expression preceding the arrow (the definiendum), the following convention is to be understood:

> $\mathbf{a}$ stands for the first variable in alphabetical order not otherwise appearing in the definiens, $\mathbf{b}$ stands for the second such variable in alphabetical order, $\mathbf{c}$ the third, and so on.

As examples, we make at once the following schematic definitions:

$$[\mathbf{M}+\mathbf{N}] \rightarrow (\lambda \mathbf{a}(\lambda \mathbf{b}((\mathbf{M}\mathbf{a})((\mathbf{N}\mathbf{a})\mathbf{b})))).$$
$$[\mathbf{M}\times\mathbf{N}] \rightarrow (\lambda \mathbf{a}(\mathbf{M}(\mathbf{N}\mathbf{a}))).$$
$$[\mathbf{M}^{\mathbf{N}}] \rightarrow (\mathbf{N}\mathbf{M}).$$

The first of these definitions means that, for instance, $[x+y]$ will be used as an abbreviation for $(\lambda a(\lambda b((x a)((y a)b))))$, and $[a+c]$ will be used as an abbreviation for $(\lambda b(\lambda d((a b)((c b)d))))$, and $[I+I]$ as an abbreviation for $(\lambda b(\lambda c((I b)((I b)c))))$, etc.

As a further device of abbreviation, we shall allow the omission of the parentheses ( ) in $(\mathbf{F}\mathbf{A})$ when this may be done without ambiguity, whether $(\mathbf{F}\mathbf{A})$ is the entire formula being written or merely some part of it. In restoring such omitted parentheses, the convention is to be followed that association is to the left (cf. Schönfinkel [49], Curry [17]). For example, $fxy$ is an abbreviation of $((fx)y)$, $f(xy)$ is an abbreviation of $(f(xy))$, $fxyz$ is an abbreviation of $(((fx)y)z)$, $f(xy)z$ is an abbreviation of $((f(xy))z)$, $f(\lambda xx)y$ is an abbreviation of $((f(\lambda xx))y)$, etc.

In expressions which (in consequence of schematic definitions) contain brackets [ ], we allow a similar omission of brackets, subject to a similar convention of association to the left; thus $x+y+z$ is an abbreviation for $[[x+y]+z]$, which expression is in turn an abbreviation for a certain well-formed formula in accordance with the schematic definition already introduced. Moreover we allow, as an abbreviation, omitting a pair of brackets and at the same time putting a dot or period in the place of the initial bracket [ ; in this case the convention, instead of association to the left, is that the omitted bracket extends from the bold period as far to the right as possible, consistently with the formula's being well-formed -- so that, for instance, $x+.y+z$ is an abbreviation for $[x+[y+z]]$, and $x+.y+.z+t$ is an abbreviation for $[x+[y+[z+t]]]$, and $(\lambda x.x+x)$ is an abbreviation for $(\lambda x[x+x])$.

We also introduce the following schematic definitions:

$$(\lambda \mathbf{x}.\mathbf{FA}) \to (\lambda \mathbf{x}(\mathbf{FA})),$$
$$(\lambda \mathbf{x}\mathbf{y}.\mathbf{FA}) \to (\lambda \mathbf{x}(\lambda \mathbf{y}(\mathbf{FA}))),$$
$$(\lambda \mathbf{x}\mathbf{y}\mathbf{z}.\mathbf{FA}) \to (\lambda \mathbf{x}(\lambda \mathbf{y}(\lambda \mathbf{z}(\mathbf{FA})))),$$

and so on for any number of variables $\mathbf{x}$, $\mathbf{y}$, $\mathbf{z}$, ... (which must be all different). And we allow similar omissions of λ's, preceding a bold period which represents an omitted bracket in the way described in the preceding paragraph -- using, e.g., $\lambda xyz.x+y+z$ as an abbreviation for $(\lambda x(\lambda y(\lambda z[[x+y]+z])))$.

Finally, we allow omission of the outside parentheses in $(\lambda \mathbf{x}\mathbf{M})$, or in $(\lambda \mathbf{x}.\mathbf{FA})$, or $(\lambda \mathbf{x}\mathbf{y}.\mathbf{FA})$, or $(\lambda \mathbf{x}\mathbf{y}\mathbf{z}.\mathbf{FA})$, etc., when this is the entire formula being written -- but not when one of these expressions appears as a proper part of a formula.

Hereafter, in writing definitions, we shall abbreviate the definiens in accordance with previously introduced abbreviations and definitions. Thus the definition of $[\mathbf{M}+\mathbf{N}]$ would now be written:

$$[M + N] \rightarrow \lambda ab \cdot Ma(Nab).$$

Definitions and other abbreviations are introduced merely as matters of convenience and are not properly part of the formal system at all. When we speak of the free variables of a formula, the bound variables of a formula, the length (number of symbols) of a formula, the occurrences of one formula as a part of another, etc., the reference is always to the unabbreviated form of the formulas in question.

The introduction and use of definitions and other abbreviations is, of course, subject to the restriction that there shall never be any ambiguity as to what formula a given abbreviated form stands for. In practice certain further restrictions are also desirable, e.g., that all free variables of the definiens be represented explicitly in the definiendum. Exact formulation of these restrictions is unnecessary for our present purpose, since all definitions and abbreviations are extraneous to the formal system, as just explained, and in principle dispensable.

6. CONVERSION. We introduce now the three following operations, or transformation rules, on well-formed formulas:

I. To replace any part $M$ of a formula by $S_y^x M|$, provided that $x$ is not a free variable of $M$ and $y$ does not occur in $M$.

II. To replace any part $((\lambda x M) N)$ of a formula by $S_N^x M|$, provided that the bound variables of $M$ are distinct both from $x$ and from the free variables of $N$.

III. To replace any part $S_N^x M|$ of a formula by $((\lambda x M) N)$, provided that $((\lambda x M) N)$ is well-formed and the bound variables of $M$ are distinct both from $x$ and from the free variables of $N$.

In the statement of these rules -- and hereafter generally -- it is to be understood that the word <u>part</u> (of a formula) means consecutive well-formed part not immediately following an occurrence of the symbol λ.

When the same formula occurs several times as such a part of another formula, each occurrence is to be counted as a different part. Thus, for instance, Rule I may be used to transform $ab(\lambda aa)(\lambda aa)$ into $ab(\lambda bb)(\lambda aa)$. Rule III may be used to transform $\lambda aa$ into $\lambda a.(\lambda aa)a$. But Rule III may not be used to transform $(\lambda aa)$ into $(\lambda((\lambda aa)a)a)$ -- the latter formula is, in fact, not even well-formed.

Rules I-III have the important property that they are effective or "definite," i.e., there is a means of always determining of any two formulas **A** and **B** whether **A** can be transformed into **B** by an application of one of the rules (and, if so, of which one).

If **A** can be transformed into **B** by an application of one of the Rules I-III, we shall say that **A** is immediately convertible into **B** (abbreviation, "**A** imc **B**"). If there is a finite sequence of formulas, in which **A** is the first formula and **B** the last, and in which each formula except the last is immediately convertible into the next one, we shall say that **A** is convertible into **B** (abbreviation, "**A** conv **B**"); and the process of obtaining **B** from **A** by a particular finite sequence of applications of Rules I-III will be called a conversion of **A** into **B** (no reference is intended to conversion in the sense of forming the converse -- for the corresponding noun we use, not "converse," but "convert"). It is not excluded that the number of applications of Rules I-III in a conversion of **A** into **B** should be zero, **B** being then the same formula as **A**.

The relation which holds between **A** and **B** when **A** conv **B** will be called interconvertibility, and we shall use the expression "**A** and **B** are interconvertible" as synonymous with "**A** conv **B**." The relation of interconvertibility is transitive, symmetric, and reflexive -- symmetric because Rules II and III are inverses of each other and Rule I is its own inverse.

If there is a conversion of **A** into **B** which contains no application of Rule II or Rule III, we shall say that **A** is convertible-I into **B** (**A** conv-I **B**). Similarly we define "**A** conv-I-II **B**" and "**A** conv-I-III **B**."

A conversion which contains no application of Rule II and

exactly one application of Rule III will be called an expansion. A conversion which contains no application of Rule III and exactly one application of Rule II will be called a reduction. If there is a reduction of **A** into **B**, we shall say that **A** is immediately reducible to **B** (**A** imr **B**). If there is a conversion of **A** into **B** which consists of one or more successive reductions, we shall say that **A** is reducible to **B** (**A** red **B**). (The meaning of "**A** red **B**" thus differs from that of "**A** conv-I-II **B**" only in that the former implies the presence of at least one application of Rule II in the conversion of **A** into **B**.)

An application of Rule II to a formula will be called a contraction of the part $((\lambda x\mathbf{M})\mathbf{N})$ which is affected.

A well-formed formula will be said to be in normal form if it contains no part of the form $((\lambda x\mathbf{M})\mathbf{N})$. We shall call **B** a normal form of **A** if **B** is in normal form and **A** conv **B**. We shall say that **A** has a normal form if there is a formula **B** which is a normal form of **A**.

A well-formed formula will be said to be in principal normal form if it is in normal form, and no variable is both a bound variable and free variable of it, and the first bound variable occurring in it (in the left-to-right order of the symbols which compose the formula) is the same as the first variable in alphabetical order which is not a free variable of it, and the variables which occur in it immediately following the symbol $\lambda$ are, when taken in the order in which they occur in the formula, in alphabetical order, without repetitions, and without omissions except of variables which are free variables of the formula. For example, $\lambda ab.ba$, and $\lambda a.a(\lambda c.bc)$, and $\lambda b.ba$ are in principal normal form; and $\lambda ac.ca$, and $\lambda bc.cb$, and $\lambda a.a(\lambda a.ba)$ are in normal form but not in principal normal form.

We shall call **B** a principal normal form of **A** if **B** is in principal normal form and **A** conv **B**. A formula in normal form is always convertible-I into a corresponding formula in principal normal form, and hence every formula which has a normal form has a principal normal form. We shall show in the next section that the principal normal form of a formula, if it exists, is unique.

An example of a formula which has no normal form (and therefore no principal normal form) is $(\lambda x.xxx)(\lambda x.xxx)$.

It is intended that, in any interpretation of the formal

calculus, only those well-formed formulas which have a normal form shall be meaningful, and, among these, interconvertible formulas shall have the same meaning. The condition of being well-formed is thus a necessary condition for meaningfulness but not a sufficient condition.

It is important that the condition of being well-formed is effective in the sense explained at the beginning of this section, whereas the condition of being well-formed and having a normal form is not effective.

7. FUNDAMENTAL THEOREMS ON WELL-FORMED FORMULAS AND ON THE NORMAL FORM. The following theorems are taken from Kleene [34] (with non-essential changes to adapt them to the present modified notation). Their proof is left to the reader; or an outline of the proof may be found in Kleene, loc. cit.

7 I. In a well-formed formula **K** there exists a unique pairing of the occurrences of the symbol (, each with a corresponding occurrence of the symbol ), in such a way that two portions of **K**, each lying between an occurrence of ( and the corresponding occurrence of ) inclusively, either are non-overlapping or else are contained one entirely within the other. Moreover, if such a pairing exists in the portion of **K** lying between the nth and the (n+r)th symbol of **K** inclusively, it is a part of the pairing in **K**. .

7 II. A necessary and sufficient condition that the portion **N** of a well-formed formula **K** which lies between a given occurrence of ( in **K** and a given occurrence of ) in **K** inclusively be well-formed is that the given occurrence of ( and the given occurrence of ) correspond.

7 III. Every well-formed formula has one of the three forms, **x**, where **x** is a variable, or (**FA**), where **F** and **A** are well-formed, or (λ**xM**), where **M** is well-formed and **x** is a free variable of **M**.

7 IV. If (**FA**) and either **F** or **A** is well-formed, then both **F** and **A** are well-formed.

7 V. If $(\lambda x M)$ is well-formed, $x$ being a variable, then $M$ is well-formed and $x$ is a free variable of $M$.

7 VI. A well-formed formula can be of the form $(FA)$, where $F$ (or $A$) is well-formed, in only one way.

7 VII. A well-formed formula can be of the form $(\lambda x M)$, where $x$ is a variable, in only one way.

7 VIII. If $P$ and $Q$ are well-formed parts of a well-formed formula $K$, then either $P$ is a part of $Q$, or $Q$ is a part of $P$, or $P$ and $Q$ are non-overlapping.

7 IX. Two distinct occurrences of the same well-formed formula $P$ as a part of a well-formed formula $K$ must be non-overlapping.

7 X. If $P$, $F$, and $A$ are well-formed and $P$ is a part of $(FA)$, then $P$ is $(FA)$ or $P$ is a part of $F$ or $P$ is a part of $A$.

7 XI. If $P$ and $M$ are well-formed and $x$ is a variable and $P$ is a part of $(\lambda x M)$, then $P$ is $(\lambda x M)$ [or $P$ is $x$] or $P$ is a part of $M$. (The clause in brackets is superfluous because of the meaning we give to the word <u>part</u> of a formula -- see §6).

7 XII. An occurrence of a variable $x$ in a well-formed formula $K$ is bound or free according as it is or is not an occurrence in a well-formed part of $K$ of the form $(\lambda x M)$. (Hence, in particular, no occurrence of a variable in a well-formed formula is both bound and free.)

7 XIII. If $M$ is well-formed and the variable $x$ is not a free variable of $M$ and the variable $y$ does not occur in $M$, then $S^x_y M|$ is well-formed and has the same free variables as $M$.

7 XIV. If $M$ and $N$ are well-formed and the variable $x$ occurs in $M$ and the bound variables of $M$ are distinct both from $x$ and from the free variables of $N$, then $S^x_N M|$ and $((\lambda x M)N)$ are well-formed and have the same free variables.

7 XV. If $K$, $P$, $Q$ are well-formed and all free variables of $P$ are also free variables of $Q$, the formula obtained

by substituting $\mathbf{Q}$ for a particular occurrence of $\mathbf{P}$ in $\mathbf{K}$, not immediately following an occurrence of $\lambda$, is well-formed.

7 XVI. If $\mathbf{A}$ is well-formed and $\mathbf{A}$ conv $\mathbf{B}$, then $\mathbf{B}$ is well-formed.

7 XVII. If $\mathbf{A}$ is well-formed and $\mathbf{A}$ conv $\mathbf{B}$, then $\mathbf{A}$ and $\mathbf{B}$ have the same free variables.

7 XVIII. If $\mathbf{K}$, $\mathbf{P}$, $\mathbf{Q}$ are well-formed, and $\mathbf{P}$ conv $\mathbf{Q}$, and $\mathbf{L}$ is obtained by substituting $\mathbf{Q}$ for a particular occurrence of $\mathbf{P}$ in $\mathbf{K}$, not immediately following an occurrence of $\lambda$, then $\mathbf{K}$ conv $\mathbf{L}$.

We shall call a well-formed part $\mathbf{P}$ of a well-formed formula $\mathbf{K}$ a free occurrence of $\mathbf{P}$ in $\mathbf{K}$ if every free occurrence of a variable in $\mathbf{P}$ is also a free occurrence of that variable in $\mathbf{K}$; in the contrary case (if some free occurrence of a variable in $\mathbf{P}$ is at the same time a bound occurrence of that variable in $\mathbf{K}$) we shall call the part $\mathbf{P}$ of $\mathbf{K}$ a bound occurrence of $\mathbf{P}$ in $\mathbf{K}$. If $\mathbf{P}$ is an occurrence of a variable in $\mathbf{K}$, not immediately following an occurrence of $\lambda$, this definition is in agreement with our previous definition of free and bound occurrences of variables.

Moreover we shall extend the notation $S^{\mathbf{x}}_{\mathbf{N}}\mathbf{M}|$ introduced in §5 by allowing $S^{\mathbf{P}}_{\mathbf{N}}\mathbf{M}|$ to stand for the result of substituting $\mathbf{N}$ for $\mathbf{P}$ throughout $\mathbf{M}$, where $\mathbf{N}$, $\mathbf{P}$, $\mathbf{M}$ are any well-formed formulas. This is possible without ambiguity, by 7 IX.

7 XIX. A well-formed part $\mathbf{P}$ of a well-formed formula $\mathbf{K}$ is a bound or free occurrence of $\mathbf{P}$ in $\mathbf{K}$ according as it is or is not an occurrence in a well-formed part of $\mathbf{K}$ of the form $(\lambda \mathbf{x}\mathbf{M})$ where $\mathbf{x}$ is a free variable of $\mathbf{P}$.

7 XX. If $\mathbf{K}$, $\mathbf{P}$, $\mathbf{Q}$ are well-formed, the formula obtained by substituting $\mathbf{Q}$ for a particular free occurrence of $\mathbf{P}$ in $\mathbf{K}$ is well-formed.

7 XXI. If $\mathbf{K}$, $\mathbf{P}$, $\mathbf{Q}$ are well-formed and there is no bound occurrence of $\mathbf{P}$ in $\mathbf{K}$, then $S^{\mathbf{P}}_{\mathbf{Q}}\mathbf{K}|$ is well-formed.

7 XXII. Let $\mathbf{x}$ be a free variable of the well-formed formula

> $\mathbf{M}$ and let $\mathbf{P}$ be the formula obtained by substituting $\mathbf{N}$ for the free occurrences of $\mathbf{x}$ in $\mathbf{M}$. If the resulting occurrences of $\mathbf{N}$ in $\mathbf{P}$ are free, $((\lambda\mathbf{x}\mathbf{M})\mathbf{N})$ conv $\mathbf{P}$.

In what follows we shall frequently make tacit assumption of these theorems.

In stating these theorems, it has been necessary to hold in abeyance the convention that formulas represented by bold capital letters are well-formed. Hereafter this convention will be restored, and formulas so represented are to be taken always as well-formed.

We turn now to a group of theorems on conversion taken from Church and Rosser [16]. In order to state these, it is necessary first to define the notion of the <u>residuals</u> of a set of parts $((\lambda\mathbf{x}_j\mathbf{M}_j)\mathbf{N}_j)$ of a formula $\mathbf{A}$ after a sequence of applications of Rules I and II to $\mathbf{A}$ (§6).

We assume that, if $p \neq q$, then $((\lambda\mathbf{x}_p\mathbf{M}_p)\mathbf{N}_p)$ is not the same part of $\mathbf{A}$ as $((\lambda\mathbf{x}_q\mathbf{M}_q)\mathbf{N}_q)$ -- though it may be the same formula. The parts $((\lambda\mathbf{x}_j\mathbf{M}_j)\mathbf{N}_j)$ of $\mathbf{A}$ need not be all the parts of $\mathbf{A}$ which have the form $((\lambda\mathbf{y}\mathbf{P})\mathbf{Q})$. The <u>residuals</u> of the $((\lambda\mathbf{x}_j\mathbf{M}_j)\mathbf{N}_j)$ after a particular sequence of applications of Rules I and II to $\mathbf{A}$ are then certain parts, of the form $((\lambda\mathbf{y}\mathbf{P})\mathbf{Q})$, of the formula into which $\mathbf{A}$ is converted by this sequence of applications of Rules I and II. They are defined as follows:

If the sequence of applications of Rules I and II in question is vacuous, each part $((\lambda\mathbf{x}_j\mathbf{M}_j)\mathbf{N}_j)$ is its own residual.

If the sequence consists of a single application of Rule I, each part $((\lambda\mathbf{x}_j\mathbf{M}_j)\mathbf{N}_j)$ is changed into a part $((\lambda\mathbf{y}_j\mathbf{M}'_j)\mathbf{N}'_j)$ of the resulting formula, and this part $((\lambda\mathbf{y}_j\mathbf{M}'_j)\mathbf{N}'_j)$ is the residual of $((\lambda\mathbf{x}_j\mathbf{M}_j)\mathbf{N}_j)$.

If the sequence consists of a single application of Rule II, let $((\lambda\mathbf{x}\mathbf{M})\mathbf{N})$ be the part of $\mathbf{A}$ which is contracted (§6), and let $\mathbf{A}'$ be the resulting formula into which $\mathbf{A}$ is converted. Let $((\lambda\mathbf{x}_p\mathbf{M}_p)\mathbf{N}_p)$ be a particular one of the $((\lambda\mathbf{x}_j\mathbf{M}_j)\mathbf{N}_j)$, and distinguish the six following cases.

Case 1: $((\lambda\mathbf{x}\mathbf{M})\mathbf{N})$ and $((\lambda\mathbf{x}_p\mathbf{M}_p)\mathbf{N}_p)$ do not overlap. Under the reduction of $A$ to $A'$, $((\lambda x_p M_p) N_p)$ goes into a definite part of $A'$, which is the same formula as $((\lambda x_p M_p) N_p)$. This part of $A'$ is the residual of $((\lambda x_p M_p) N_p)$.

Case 2: $((\lambda x M) N)$ is a part of $M_p$. Under the reduction of $A$ to $A'$, $M_p$ goes into a definite part $M_p'$ of $A$, which arises from $M_p$ by contraction of $((\lambda x M) N)$, and $((\lambda x_p M_p) N_p)$ goes into the part $((\lambda x_p M_p') N_p)$ of $A'$. This part $((\lambda x_p M_p') N_p)$ of $A'$ is the residual of $((\lambda x_p M_p) N_p)$.

Case 3: $((\lambda x M) N)$ is a part of $N_p$. Under the reduction of $A$ to $A'$, $N_p$ goes into a definite part $N_p'$ of $A'$, which arises from $N_p$ by contraction of $((\lambda x M) N)$, and $((\lambda x_p M_p) N_p)$ goes into the part $((\lambda x_p M_p) N_p')$ of $A'$. This part $((\lambda x_p M_p) N_p')$ of $A'$ is the residual of $((\lambda x_p M_p) N_p)$.

Case 4: $((\lambda x M) N)$ is $((\lambda x_p M_p) N_p)$. In this case $((\lambda x_p M_p) N_p)$ has no residual in $A'$.

Case 5: $((\lambda x_p M_p) N_p)$ is a part of $M$. Let $M'$ be the result of replacing all $x$'s of $M$ except those occurring in $((\lambda x_p M_p) N_p)$ by $N$. Under these changes the part $((\lambda x_p M_p) N_p)$ of $M$ goes into a definite part of $M'$ which we shall denote also by $((\lambda x_p M_p) N_p)$, since it is the same formula. If now we replace $((\lambda x_p M_p) N_p)$ in $M'$ by $S^x_N((\lambda x_p M_p) N_p)|$, $M'$ becomes $S^x_N M|$ and we denote by $S^x_N((\lambda x_p M_p) N_p)|$ the particular occurrence of $S^x_N((\lambda x_p M_p) N_p)|$ in $S^x_N M|$ that resulted from replacing $((\lambda x_p M_p) N_p)$ in $M'$ by the formula $S^x_N((\lambda x_p M_p) N_p)|$. Then the residual in $A'$ of $((\lambda x_p M_p) N_p)$ in $A$ is defined to be the part $S^x_N((\lambda x_p M_p) N_p)|$ in the particular occurrence of $S^x_N M|$ in $A'$ that resulted from replacing $((\lambda x M) N)$ in $A$ by $S^x_N M|$.

Case 6: $((\lambda x_p M_p) N_p)$ is a part of $N$. Let $((\lambda y_1 P_1) Q_1)$ respectively stand for the particular occurrences of the formula $((\lambda x_p M_p) N_p)$ in $S^x_N M|$ which are the part $((\lambda x_p M_p) N_p)$ in each of those particular occurrences of the formula $N$ in $S^x_N M|$ that resulted from replacing the $x$'s of $M$ by $N$. Then the residuals in $A'$ of $((\lambda x_p M_p) N_p)$ in $A$ are the parts $((\lambda y_1 P_1) Q_1)$ in the particular occurrence of the formula $S^x_N M|$ in $A'$ that resulted from replacing $((\lambda x M) N)$ in $A$ by $S^x_N M|$.

Finally, in the case of a sequence of two or more successive applications of Rules I, II to $\mathbf{A}$, say $\mathbf{A}$ imc $\mathbf{A}'$ imc $\mathbf{A}''$ imc ... , we define the residuals in $\mathbf{A}'$ of the parts $((\lambda x_j \mathbf{M}_j)\mathbf{N}_j)$ of $\mathbf{A}$ in the way just described, and we define the residuals in $\mathbf{A}''$ of the parts $((\lambda x_j \mathbf{M}_j)\mathbf{N}_j)$ of $\mathbf{A}$ to be the residuals of the residuals in $\mathbf{A}'$, and so on.

7 XXIII. After a sequence of applications of Rules I and II to $\mathbf{A}$, under which $\mathbf{A}$ is converted into $\mathbf{B}$, the residuals of the parts $((\lambda x_j \mathbf{M}_j)\mathbf{N}_j)$ of $\mathbf{A}$ are a set (possibly vacuous) of parts of $\mathbf{B}$ which each have the form $((\lambda \mathbf{y}\mathbf{P})\mathbf{Q})$.

7 XXIV. After a sequence of applications of Rules I and II to $\mathbf{A}$, no residual of the part $((\lambda \mathbf{x}\mathbf{M})\mathbf{N})$ of $\mathbf{A}$ can coincide with a residual of the part $((\lambda \mathbf{x}'\mathbf{M}')\mathbf{N}')$ of $\mathbf{A}$ unless $((\lambda \mathbf{x}\mathbf{M})\mathbf{N})$ coincides with $((\lambda \mathbf{x}'\mathbf{M}')\mathbf{N}')$.

We say that a sequence of reductions on $\mathbf{A}_1$, say $\mathbf{A}_1$ imr $\mathbf{A}_2$ imr $\mathbf{A}_3$ ... imr $\mathbf{A}_{n+1}$, is a <u>sequence of contractions on the parts</u> $((\lambda x_j \mathbf{M}_j)\mathbf{N}_j)$ <u>of</u> $\mathbf{A}_1$ if the reduction from $\mathbf{A}_i$ to $\mathbf{A}_{i+1}$ $(i = 1, \ldots, n)$ involves a contraction of a residual of the $((\lambda x_j \mathbf{M}_j)\mathbf{N}_j)$. Moreover, if no residuals of the $((\lambda x_j \mathbf{M}_j)\mathbf{N}_j)$ occur in $\mathbf{A}_{n+1}$ we say that the sequence of contractions on the $((\lambda x_j \mathbf{M}_j)\mathbf{N}_j)$ <u>terminates</u> and that $\mathbf{A}_{n+1}$ is the result.

In some cases we wish to speak of a sequence of contractions on the parts $((\lambda x_j \mathbf{M}_j)\mathbf{N}_j)$ of $\mathbf{A}$ where the set $((\lambda x_j \mathbf{M}_j)\mathbf{N}_j)$ may be vacuous. To handle this we agree that, if the set $((\lambda x_j \mathbf{M}_j)\mathbf{N}_j)$ is vacuous, the sequence of contractions shall be a vacuous sequence of reductions.

7 XXV. If $((\lambda x_j \mathbf{M}_j)\mathbf{N}_j)$ are parts of $\mathbf{A}$, then a number m can be found such that any sequence of contractions on the $((\lambda x_j \mathbf{M}_j)\mathbf{N}_j)$ will terminate after at most m contractions, and if $\mathbf{A}'$ and $\mathbf{A}''$ are two results of terminating sequences of contractions on the $((\lambda x_j \mathbf{M}_j)\mathbf{N}_j)$, then $\mathbf{A}'$ conv-I $\mathbf{A}''$.

This is proved by induction on the length of $\mathbf{A}$. It is trivially true if the length of $\mathbf{A}$ is 1 (i.e., if $\mathbf{A}$ consists of a single symbol), the number m being then 0. As hypothesis of induction, assume that the proposition is true of every formula $\mathbf{A}$ of length less than n. On this hypothesis we have to prove that the proposition is true of an arbitrary given formula $\mathbf{A}$ of length n. This we proceed to do, by means of a proof involving three cases.

Case 1: $\mathbf{A}$ has the form $\lambda \mathbf{x}\mathbf{M}$. All the parts $((\lambda \mathbf{x}_j \mathbf{M}_j)\mathbf{N}_j)$ of $\mathbf{A}$ must be parts of $\mathbf{M}$. Since $\mathbf{M}$ is of length less than n, we apply the hypothesis of induction to $\mathbf{M}$.

Case 2: $\mathbf{A}$ has the form $\mathbf{F}\mathbf{X}$, where $\mathbf{F}\mathbf{X}$ is not one of the $((\lambda \mathbf{x}_j \mathbf{M}_j)\mathbf{N}_j)$. All the parts $((\lambda \mathbf{x}_j \mathbf{M}_j)\mathbf{N}_j)$ of $\mathbf{A}$ must be parts either of $\mathbf{F}$ or of $\mathbf{X}$. Since $\mathbf{F}$ and $\mathbf{X}$ are each of length less than n, we apply the hypothesis of induction.

Case 3: $\mathbf{A}$ is $((\lambda \mathbf{x}_p \mathbf{M}_p)\mathbf{N}_p)$, where $((\lambda \mathbf{x}_p \mathbf{M}_p)\mathbf{N}_p)$ is one of the $((\lambda \mathbf{x}_j \mathbf{M}_j)\mathbf{N}_j)$. By the hypothesis of induction, there is a number a such that any sequence of contractions on those $((\lambda \mathbf{x}_j \mathbf{M}_j)\mathbf{N}_j)$ which are parts of $\mathbf{M}_p$ terminates after at most a contractions, and there is a number b such that any sequence of contractions on those $((\lambda \mathbf{x}_j \mathbf{M}_j)\mathbf{N}_j)$ which are parts of $\mathbf{N}_p$ terminates after at most b contractions; moreover, if we start with the formula $\mathbf{M}_p$ and perform a terminating sequence of contractions on those $((\lambda \mathbf{x}_j \mathbf{M}_j)\mathbf{N}_j)$ which are parts of $\mathbf{M}_p$, the result is a formula $\mathbf{M}$, which is unique to within applications of Rule I, and which contains a certain number c, $\geqq 1$, of free occurrences of the variable $\mathbf{x}_p$.

Now one way of performing a terminating sequence of contractions on the parts $((\lambda \mathbf{x}_j \mathbf{M}_j)\mathbf{N}_j)$ of $\mathbf{A}$ is as follows. First perform a terminating sequence of contractions on those $((\lambda \mathbf{x}_j \mathbf{M}_j)\mathbf{N}_j)$ which are parts of $\mathbf{M}_p$, so converting $\mathbf{A}$ into $((\lambda \mathbf{t}\mathbf{M})\mathbf{N}_p)$. Then there is one and only one residual of $((\lambda \mathbf{x}_p \mathbf{M}_p)\mathbf{N}_p)$, namely the entire formula $((\lambda \mathbf{x}\mathbf{M})\mathbf{N}_p)$. Perform a contraction of this, so obtaining

$$S^{\mathbf{t}}_{\mathbf{N}_p} \mathbf{M}' |$$

where $\mathbf{M}'$ differs from $\mathbf{M}$ at most by applications of Rule I. Then in this formula there are c occurrences of $\mathbf{N}_p$ resulting from the substitution of $\mathbf{N}_p$ for $\mathbf{t}$. Take each of these occurrences of $\mathbf{N}_p$ in order and perform a terminating sequence of contractions on the residuals of the $((\lambda \mathbf{x}_j \mathbf{M}_j)\mathbf{N}_j)$ occurring in it.

Let us call such a terminating sequence of contractions on the parts $((\lambda \mathbf{x}_j \mathbf{M}_j)\mathbf{N}_j)$ of $\mathbf{A}$ a <u>special</u> terminating sequence of contractions on the parts $((\lambda \mathbf{x}_j \mathbf{M}_j)\mathbf{N}_j)$ of $\mathbf{A}$. Clearly such a special terminating sequence of contractions contains at most $a+1+cb$ contractions.

Consider now any sequence of contractions, $\mu$, on the parts $((\lambda \mathbf{x}_j \mathbf{M}_j)\mathbf{N}_j)$ of $\mathbf{A}$. The part $((\lambda \mathbf{x}_p \mathbf{M}_p)\mathbf{N}_p)$ of $\mathbf{A}$ will have just one residual (which will always be the entire formula) up to the point that a contraction of its residual occurs, and thereafter will have no residual; moreover, if the sequence of contractions is continued, a contraction of the residual of $((\lambda \mathbf{x}_p \mathbf{M}_p)\mathbf{N}_p)$ must occur within at most $a+b+1$ contractions. Hence we may suppose, without loss of generality, that $\mu$ consists of a sequence of contractions, $\phi$, on the $((\lambda \mathbf{x}_j \mathbf{M}_j)\mathbf{N}_j)$ which are different from $((\lambda \mathbf{x}_p \mathbf{M}_p)\mathbf{N}_p)$, followed by a contraction $\beta_0$ of the residual of $((\lambda \mathbf{x}_p \mathbf{M}_p)\mathbf{N}_p)$, followed by a sequence of contractions, $\vartheta$, on the then remaining residuals of the $((\lambda \mathbf{x}_j \mathbf{M}_j)\mathbf{N}_j)$. Clearly, $\phi$ can be replaced by a sequence of contractions, $\alpha_0$, on the $((\lambda \mathbf{x}_j \mathbf{M}_j)\mathbf{N}_j)$ which are parts of $\mathbf{M}_p$, followed by a sequence of contractions, $\eta$, on the $((\lambda \mathbf{x}_j \mathbf{M}_j)\mathbf{N}_j)$ which are parts of $\mathbf{N}_p$ -- in the sense that $\alpha_0$ followed by $\eta$ gives the same end formula as $\phi$ and the same set of residuals for each of the $((\lambda \mathbf{x}_j \mathbf{M}_j)\mathbf{N}_j)$. Moreover, replacing $\phi$ by $\alpha_0$ followed by $\eta$ does not change the total number of contractions of residuals of parts of $\mathbf{M}_p$ or of residuals of parts of $\mathbf{N}_p$. Next, $\eta$ followed by $\beta_0$ can be replaced by a contraction $\beta'$ of the residual $((\lambda \mathbf{y} \mathbf{P})\mathbf{N}_p)$ of $((\lambda \mathbf{x}_p \mathbf{M}_p)\mathbf{N}_p)$ followed by a set of applications of $\eta$ on each of those occurrences of $\mathbf{N}_p$ in the resulting formula

$$S^{\mathbf{y}}_{\mathbf{N}_p} \mathbf{P}' |$$

which arose by substituting $\mathbf{N}_p$ for $\mathbf{y}$ in $\mathbf{P}'$. (Here $\mathbf{P}'$ differs from $\mathbf{P}$ at most by applications of Rule I. Since $\eta$ may be thought of as a transformation of the formula $\mathbf{N}_p$, the convention will be understood which we use when we speak of the sequence of reductions of a given formula which results from applying $\eta$ to a particular occurrence of $N_p$ in that formula.)

By this means the sequence of contractions, $\mu$, is replaced by a sequence of contractions, $\mu'$, which consists of a sequence of contractions, $\alpha_0$, on the $((\lambda x_j M_j)N_j)$ which are parts of $M_p$, followed by a contraction $\beta'$ of the residual of $((\lambda x_p M_p)N_p)$, followed by further contractions on the then remaining residuals of the $((\lambda x_j M_j)N_j)$.

Consider now the part $\zeta$ of $\mu'$, consisting of $\beta'$ and the contractions that follow it, up to and including the first contraction of a residual of a part of $M_p$. Denoting the formula on which $\zeta$ acts by $((\lambda y P)N_p)$, we see that $\zeta$ can be considered as the act of first replacing the free $y$'s of $P'$ by various formulas $N_{pk}$, got from $N_p$ by various sequences of reductions (which may be vacuous), and then (possibly after some applications of Rule I) contracting a residual $((\lambda z R)S)$ of one of the $((\lambda x_j M_j)N_j)$ which are parts of $M_p$, say $((\lambda x_q M_q)N_q)$. From this point of view, we see that none of the free $z$'s of $R$ are parts of any $N_{pk}$, and hence $\zeta$ can be replaced by a contraction (possibly after some applications of Rule I) of that residual in $P$ of $((\lambda x_q M_q)N_q)$ of which $((\lambda z R) S)$ is a residual, followed by a contraction (possibly after some applications of Rule I) of the residual of $((\lambda x_p M_p)N_p)$, followed by a sequence of contractions on residuals of parts of $N_p$.

If $\mu'$ is altered by replacing $\zeta$ in this way, the result is a sequence of contractions, $\mu''$, having the same form as $\mu'$, but having the property that after the contraction of the residual of $((\lambda x_p M_p)N_p)$ one less contraction of residuals of parts of $M_p$ occurs.

By repetitions of this process, $\mu$ is finally replaced by a sequence of contractions $\nu$, which consists of a sequence of contractions, $\alpha$, on the $((\lambda x_j M_j)N_j)$ which are parts of $M_p$, followed by a contraction $\beta$ of the residual of $((\lambda x_p M_p)N_p)$, followed by a sequence of contractions $\gamma$ on residuals of the $((\lambda x_j M_j)N_j)$ which are parts of $N_p$. Moreover, $\nu$ contains at least as many contractions as $\mu$ -- for in the process of obtaining $\nu$ from $\mu$ there is no step which can decrease the number of contractions. The sequence of contractions, $\alpha$, contains at most a contractions, and $\gamma$ contains at most cb contractions. Thus $\nu$, and consequently $\mu$, contains at most a+1+cb contractions.

Thus we have proved that any sequence of contractions on the parts $((\lambda \mathbf{x}_j \mathbf{M}_j)\mathbf{N}_j)$ of $\mathbf{A}$ will terminate after at most a+1+cb contractions.

Now suppose that $\mu$ is a <u>terminating</u> sequence of contractions. Then $\nu$ either is a <u>special</u> terminating sequence of contractions (see above) or can be made so by some evident changes in the order in which the contractions in $\gamma$ are performed. By the hypothesis of induction, applied to $\mathbf{M}_p$ and $\mathbf{N}_p$, the result of a special terminating sequence of contractions is unique to within possible applications of Rule I. Therefore the result of any terminating sequence of contractions, $\mu$, is unique to within possible applications of Rule I.

7 XXVI. If $\mathbf{A}$ imr $\mathbf{B}$ by a contraction of the part $((\lambda \mathbf{x}\mathbf{M})\mathbf{N})$ of $\mathbf{A}$, and $\mathbf{A}_1$ is $\mathbf{A}$, and $\mathbf{A}_1$ imr $\mathbf{A}_2$, $\mathbf{A}_2$ imr $\mathbf{A}_3$, ..., and, for all k, $\mathbf{B}_k$ is the result of a terminating sequence of contractions on the residuals in $\mathbf{A}_k$ of $((\lambda \mathbf{x}\mathbf{M})\mathbf{N})$, then:

(1) $\mathbf{B}_1$ is $\mathbf{B}$.

(2) For all k, $\mathbf{B}_k$ conv-I-II $\mathbf{B}_{k+1}$:

(3) Even if the sequence $\mathbf{A}_1$, $\mathbf{A}_2$, ... can be continued to infinity, there is a number $u_m$, depending on the formula $\mathbf{A}$, the part $((\lambda \mathbf{x}\mathbf{M})\mathbf{N})$ of $\mathbf{A}$, and the number m, such that, starting with $\mathbf{B}_m$, at most $u_m$ consecutive $\mathbf{B}_k$'s occur for which it is not true that $\mathbf{B}_k$ red $\mathbf{B}_{k+1}$.

(1) is obvious.

To prove (2), let $((\lambda \mathbf{y}_1 \mathbf{P}_1)\mathbf{Q}_1)$ be the residuals in $\mathbf{A}_k$ of $((\lambda \mathbf{x}\mathbf{M})\mathbf{N})$ and let the reduction of $\mathbf{A}_k$ into $\mathbf{A}_{k+1}$ involve a contraction of (a residual of) the part $((\lambda \mathbf{z}\mathbf{R})\mathbf{S})$ of $\mathbf{A}_k$. Then $\mathbf{B}_{k+1}$ is the result of a terminating sequence of contractions on $((\lambda \mathbf{z}\mathbf{R})\mathbf{S})$ and the parts $((\lambda \mathbf{y}_1 \mathbf{P}_1)\mathbf{Q}_1)$ of $\mathbf{A}_k$. If $((\lambda \mathbf{z}\mathbf{R})\mathbf{S})$ is one of the $((\lambda \mathbf{y}_1 \mathbf{P}_1)\mathbf{Q}_1)$, no residuals of $((\lambda \mathbf{z}\mathbf{R})\mathbf{S})$ occur in $\mathbf{B}_k$, and $\mathbf{B}_k$ conv-I $\mathbf{B}_{k+1}$ by 7 XXV. If, however, $((\lambda \mathbf{z}\mathbf{R})\mathbf{S})$ is not one of the $((\lambda \mathbf{y}_1 \mathbf{P}_1)\mathbf{Q}_1)$, a set of residuals of $((\lambda \mathbf{z}\mathbf{R})\mathbf{S})$

does occur in $\boldsymbol{B}_k$ and a terminating sequence of contractions on these residuals in $\boldsymbol{B}_k$ gives $\boldsymbol{B}_{k+1}$ by 7 XXV.

Thus $\boldsymbol{B}_k$ red $\boldsymbol{B}_{k+1}$ unless the reduction of $\boldsymbol{A}_k$ into $\boldsymbol{A}_{k+1}$ involves a contraction of a residual of $((\lambda \boldsymbol{x}\boldsymbol{M})\boldsymbol{N})$; but if we start with any particular $\boldsymbol{A}_k$ this can be the case only a finite number of successive times by 7 XXV. Hence (3) is proved, $u_m$ being defined as follows:

Perform m successive reductions on $\boldsymbol{A}$ in all possible ways. This gives a finite set of formulas (since, for this purpose, we need not distinguish formulas differing only by applications of Rule I). In each formula find the largest number of reductions that can occur in a terminating sequence of contractions on the residuals of $((\lambda \boldsymbol{x}\boldsymbol{M})\boldsymbol{N})$. Then let $u_m$ be the largest of these.

7 XXVII. If $\boldsymbol{A}$ conv $\boldsymbol{B}$, there is a conversion of $\boldsymbol{A}$ into $\boldsymbol{B}$ in which no expansion precedes any reduction.

In the given conversion of $\boldsymbol{A}$ into $\boldsymbol{B}$, let the last expansion which precedes any reduction be an expansion of $\boldsymbol{B}_1$ into $\boldsymbol{A}_1$. This expansion is followed by a sequence of one or more reductions, say $\boldsymbol{A}_1$ imr $\boldsymbol{A}_2$, $\boldsymbol{A}_2$imr $\boldsymbol{A}_3$, ..., $\boldsymbol{A}_{n-1}$ imr $\boldsymbol{A}_n$, and $\boldsymbol{A}_n$ conv-I-III $\boldsymbol{B}$. The inverse of the expansion of $\boldsymbol{B}_1$ into $\boldsymbol{A}_1$ is a reduction of $\boldsymbol{A}_1$ into $\boldsymbol{B}_1$; let $((\lambda \boldsymbol{x}\boldsymbol{M})\boldsymbol{N})$ be the part of $\boldsymbol{A}$ which is contracted in this reduction, and let $\boldsymbol{B}_k$ (k = 2, 3, ..., n) be the result of a terminating sequence of contractions on the residuals in $\boldsymbol{A}_k$ of $((\lambda \boldsymbol{x}\boldsymbol{M})\boldsymbol{N})$. By 7 XXVI, $\boldsymbol{B}_1$ conv-I-II $\boldsymbol{B}_2$, $\boldsymbol{B}_2$ conv-I-II $\boldsymbol{B}_3$, ..., $\boldsymbol{B}_{n-1}$ conv-I-II $\boldsymbol{B}_n$, $\boldsymbol{B}_n$ conv-I-III $\boldsymbol{A}_n$, $\boldsymbol{A}_n$ conv-I-III $\boldsymbol{B}$. This provides an alternative conversion of $\boldsymbol{B}_1$ into $\boldsymbol{B}$ in which no expansion precedes any reduction. The given conversion of $\boldsymbol{A}$ into $\boldsymbol{B}$ may be altered by employing this alternative conversion of $\boldsymbol{B}_1$ into $\boldsymbol{B}$ instead of the one originally involved, with the result that the number of expansions which are out of place (precede reductions) in the conversion of $\boldsymbol{A}$ into $\boldsymbol{B}$ is decreased by one. Repetitions of this process lead to a conversion of $\boldsymbol{A}$ into $\boldsymbol{B}$ in which no expansion precedes reductions.

7 XXVIII. If $\boldsymbol{B}$ is a normal form of $\boldsymbol{A}$, then $\boldsymbol{A}$ conv-I-II $\boldsymbol{B}$.

This is a corollary of 7 XXVII, since no reductions are possible of a formula in normal form.

7 XXIX. If $\boldsymbol{A}$ has a normal form, its normal form is unique to within applications of Rule I.

For if $\boldsymbol{B}$ and $\boldsymbol{B}'$ are both normal forms of $\boldsymbol{A}$, then $\boldsymbol{B}'$ is a normal form of $\boldsymbol{B}$. Hence $\boldsymbol{B}$ conv-I-II $\boldsymbol{B}'$. Hence $\boldsymbol{B}$ conv-I $\boldsymbol{B}$, since no reductions are possible of the normal form $\boldsymbol{B}$.

Note that 7 XXIX ensures a kind of consistency of the calculus of $\lambda$-conversion, in that certain formulas for which different interpretations are intended are shown not to be interconvertible.

7 XXX. If $\boldsymbol{A}$ has a normal form, it has a unique principal normal form.

7 XXXI. If $\boldsymbol{B}$ is a normal form of $\boldsymbol{A}$, then there is a number m such that any sequence of reductions starting from $\boldsymbol{A}$ will lead to $\boldsymbol{B}$ (to within applications of Rule I) after at most m reductions.

In order to prove 7 XXXI, we first prove the following lemma by induction on n:

> If $\boldsymbol{B}$ is a normal form of $\boldsymbol{A}$ and there is a sequence of n reductions leading from $\boldsymbol{A}$ to $\boldsymbol{B}$, then there is a number $v_{\boldsymbol{A},n}$ such that any sequence of reductions starting from $\boldsymbol{A}$ will lead to a normal form of $\boldsymbol{A}$ in at most $v_{\boldsymbol{A},n}$ reductions.

If $n = 0$, we take $v_{\boldsymbol{A},0}$ to be 0.

Assume, as hypothesis of induction, that the lemma is true when $n = k$. Suppose $\boldsymbol{A}$ imr $\boldsymbol{C}$, $\boldsymbol{C}$ imr $\boldsymbol{C}_1$, $\boldsymbol{C}_1$ imr $\boldsymbol{C}_2$, $\boldsymbol{C}_2$ imr $\boldsymbol{C}_3$, ..., $\boldsymbol{C}_{k-1}$ imr $\boldsymbol{B}$. Also, where $\boldsymbol{A}_1$ is the same as $\boldsymbol{A}$, suppose $\boldsymbol{A}_1$ imr $\boldsymbol{A}_2$, $\boldsymbol{A}_2$ imr $\boldsymbol{A}_3$, ... . By 7 XXVI there is a sequence ($\boldsymbol{D}_1$ the same as $\boldsymbol{C}$), $\boldsymbol{D}_1$ conv-I-II $\boldsymbol{D}_2$, $\boldsymbol{D}_2$ conv-I-II $\boldsymbol{D}_3$, ..., such that $\boldsymbol{A}_j$ conv-I-II $\boldsymbol{D}_j$ for all j's for which $\boldsymbol{A}_j$ exists; and, if the reduction from $\boldsymbol{A}$ to $\boldsymbol{C}$ involves a contraction of $((\lambda \boldsymbol{x}\boldsymbol{M})\boldsymbol{N})$, then, starting with $\boldsymbol{D}_m$, at most $u_m$ consecutive

$\mathbf{D}_j$'s occur for which it is not true that $\mathbf{D}_j$ red $\mathbf{D}_{j+1}$.

Since the sequence $\mathbf{C}$ imr $\mathbf{C}_1$, $\mathbf{C}_1$ imr $\mathbf{C}_2$, ... leads to $\mathbf{B}$ in k reductions, there is, by hypothesis of induction, a number $v_{\mathbf{C},k}$ such that any sequence of reductions starting from $\mathbf{C}$ leads to a normal form (and thus terminates) after at most $v_{\mathbf{C},k}$ reductions. Hence there are at most $v_{\mathbf{C},k}$ reductions in the sequence $\mathbf{D}_1$ conv-I-II $\mathbf{D}_2$, $\mathbf{D}_2$ conv-I-II $\mathbf{D}_3$, ..., and this sequence must terminate after at most $f(v_{\mathbf{C},k})$ steps, $f(x)$ being defined as follows:

$$f(0) = u_1,$$

$$f(x+1) = f(x)+M+1,$$

where M is the greatest of the numbers $u_1, u_2, \ldots, u_{f(x)+1}$. (Of course $f(x)$ depends on the formula $\mathbf{A}$ and the part $((\lambda x\mathbf{M})\mathbf{N})$ of $\mathbf{A}$, as well as on x, because $u_m$ depends on $\mathbf{A}$ and $((\lambda x\mathbf{M})\mathbf{N})$.

Since the sequence of $\mathbf{D}_j$'s continues as long as there are $\mathbf{A}_j$'s on which reductions can be performed, it follows that after at most $f(v_{\mathbf{C},k})$ reductions an $\mathbf{A}_j$ is reached on which no reductions are possible. But this is equivalent to saying that this $\mathbf{A}_j$ is in normal form. Thus any reductionsof $\mathbf{A}$ to a formula $\mathbf{C}$, such that there is a sequence of k reductions leading from $\mathbf{C}$ to a normal form of $\mathbf{A}$, determines an upper bound, $f(v_{\mathbf{C},k})$, which holds for all sequences of reductions starting from $\mathbf{A}$. Since the number of possible reductions of $\mathbf{A}$ to such formulas $\mathbf{C}$ is finite (reductions, or formulas $\mathbf{C}$, which differ only by applications of Rule I need not be distinguished as different), we take $v_{\mathbf{A},k+1}$ to be the least of the numbers $f(v_{\mathbf{C},k})$.

This completes the proof of the lemma. Hence 7 XXXI follows by 7 XXVIII.

7 XXXII. If $\mathbf{A}$ has a normal form, every [well-formed] part of $\mathbf{A}$ has a normal form.

This follows from 7 XXXI, since any sequence of reductions on a part of $\mathbf{A}$ implies a sequence of reductions on $\mathbf{A}$ and therefore must terminate.

## Chapter III

## LAMBDA-DEFINABILITY

8. LAMBDA-DEFINABILITY OF FUNCTIONS OF POSITIVE INTEGERS. We define,

$$1 \rightarrow \lambda ab.ab,$$

$$2 \rightarrow \lambda ab.a(ab),$$

$$3 \rightarrow \lambda ab.a(a(ab)),$$

and so on, each numeral (in the Arabic decimal notation) being introduced as an abbreviation for a corresponding formula of the indicated form. But where a numeral consists of more than one digit, a bar is used over it, in order to avoid confusion with other notations; thus,

$$\overline{11} \rightarrow \lambda ab.a(a(a(a(a(a(a(a(a(a(ab)))))))))),$$

but 11, without the bar, is an abbreviation for

$$(\lambda ab.ab)(\lambda ab.ab).$$

In connection with these definitions an interpretation of the calculus of $\lambda$-conversion is contemplated under which each of the formulas abbreviated as a numeral is interpreted as denoting the corresponding positive integer. Since it is intended at the same time to retain the interpretation of the formulas of the calculus (which have a normal form) as denoting certain functions in accordance with the ideas of Chapter I, this means that the positive integers are identified with certain functions. For example, the number 2 is identified with the function which, when applied to the function $f$ as argument, yields the product of $f$ by itself (product in the sense of the product, or resultant, of two transformations); similarly the number 14 is identified with the function which, when applied to the function $f$ as argument, yields the fourteenth power of $f$ (power in the sense of power of a transformation). This is allowable on the ground that abstract number theory requires of the positive integers only that they form a progression and, subject to this condition, the integers may be identified with any entities whatever; as a matter of fact, logical constructions of the positive integers by identifying them with entities thought to be logically more fundamental are possible in many different ways (the present method should be compared with that familiar in the works of Frege and Russell, according to which the non-negative integers are identified with classes of similar finite classes).

A function $F$ of positive integers -- i.e., a function of one variable for which the range of arguments and the range of values each consist of positive integers -- is said to be <u>λ-definable</u> if there is a formula $\mathbf{F}$ such that (1) whenever $m$ and $n$ are positive integers, and $Fm = n$, and $\mathbf{M}$ and $\mathbf{N}$ are the formulas which represent (denote) the integers $m$ and $n$ respectively, then $\mathbf{FM}$ conv $\mathbf{N}$, and (2) whenever the function $F$ has no value for the positive integer $m$ as argument, and $\mathbf{M}$ represents $m$, then $\mathbf{FM}$ has no normal form. Similarly the function $F$ of two integer variables is said to be <u>λ-definable</u> if there is a formula $\mathbf{F}$ such that (1) if $l$, $m$, $n$ are positive integers, and $Flm = n$, and $\mathbf{L}$, $\mathbf{M}$, $\mathbf{N}$ represent the integers $l$, $m$, $n$ respectively, then $\mathbf{FLM}$ conv $\mathbf{N}$, and (2) if the function $F$ has no value for the positive integers $l$, $m$ as arguments, and $\mathbf{L}$, $\mathbf{M}$ represent $l$, $m$ respectively, then $\mathbf{FLM}$ has no normal form. And so on, for functions of any number of variables.

We shall say also, under the circumstances described, that the formula $\mathbf{F}$ <u>λ-defines</u> the function $F$ (we use the word "λ-defines rather than "denotes" or "represents" only because the function which $\mathbf{F}$ denotes, in general has other elements than positive integers in its range -- or ranges -- of arguments).

The successor function of positive integers (i.e., the function $x+1$) is λ-defined by the formula $S$, where

$$S \longrightarrow \lambda abc.b(abc).$$

It is left to the reader to verify this, and also to verify that addition, and multiplication, and exponentiation of positive integers are $\lambda$-defined by the formulas $\lambda mn.m+n$, and $\lambda mn.m\times n$, and $\lambda mn.m^n$ respectively (see definitions in §5).

These $\lambda$-definitions of addition, multiplication, and exponentiation are due to Rosser (see Kleene [35]). The definition of multiplication depends on the observation that the product of two positive integers in the sense of the product of transformations is the same as their product in the arithmetic sense, and the definition of exponentiation then follows because, when the positive integer $n$ is taken of any function $f$ as argument, there results the $n$th power of $f$ in the sense of the product of transformations.

The reader may also verify that, for any formulas $\mathbf{L}$, $\mathbf{M}$, $\mathbf{N}$ (whether representing positive integers or not):

$$[\mathbf{L}+\mathbf{M}]+\mathbf{N} \text{ conv } \mathbf{L}+[\mathbf{M}+\mathbf{N}],$$

$$[\mathbf{L}\times\mathbf{M}]\times\mathbf{N} \text{ conv } \mathbf{L}\times[\mathbf{M}\times\mathbf{N}],$$

$$[\mathbf{L}+\mathbf{M}]\times\mathbf{N} \text{ conv } [\mathbf{L}\times\mathbf{N}]+[\mathbf{M}\times\mathbf{N}],$$

$$\mathbf{L}^{\mathbf{M}+\mathbf{N}} \text{ conv } \mathbf{L}^{\mathbf{M}}\times\mathbf{L}^{\mathbf{N}},$$

$$\mathbf{L}^{\mathbf{M}\times\mathbf{N}} \text{ conv } [\mathbf{L}^{\mathbf{N}}]^{\mathbf{M}},$$

$$S\mathbf{M} \text{ conv } 1+\mathbf{M}.$$

9. ORDERED PAIRS AND TRIADS, THE PREDECESSOR FUNCTION. We now introduce formulas which may be thought of as representing ordered pairs and ordered triads, as follows:

$$[\mathbf{M}, \mathbf{N}] \rightarrow \lambda a.a\mathbf{M}\mathbf{N},$$

$$[\mathbf{L}, \mathbf{M}, \mathbf{N}] \rightarrow \lambda a.a\mathbf{L}\mathbf{M}\mathbf{N},$$

$$2_1 \rightarrow \lambda a.a(\lambda bc.cIb),$$

$$2_2 \rightarrow \lambda a.a(\lambda bc.bIc),$$

$$3_1 \rightarrow \lambda a.a(\lambda bcd.cIdIb),$$

$$3_2 \rightarrow \lambda a.a(\lambda bcd.bIdIc),$$

$$3_3 \rightarrow \lambda a.a(\lambda bcd.bIcId).$$

If $\mathbf{L}$, $\mathbf{M}$, $\mathbf{N}$ are formulas representing positive integers, then $2_1[\mathbf{M}, \mathbf{N}]$ conv $\mathbf{M}$, $2_2[\mathbf{M}, \mathbf{N}]$ conv $\mathbf{N}$, $3_1[\mathbf{L}, \mathbf{M}, \mathbf{N}]$ conv $\mathbf{L}$, $3_2[\mathbf{L}, \mathbf{M}, \mathbf{N}]$ conv $\mathbf{M}$, and $3_3[\mathbf{L}, \mathbf{M}, \mathbf{N}]$ conv $\mathbf{N}$.

Verification of this depends on the observation that, if $\mathbf{M}$ is a formula representing a positive integer, $\mathbf{M}I$ conv $I$ (the $m$th power of the identity is the identity).

By the predecessor function of positive integers we mean the function whose value for the argument 1 is 1 and whose value for any other positive integer argument $x$ is $x-1$. This function is λ-defined by

$$P \rightarrow \lambda a \cdot 3_3(a(\lambda b[S(3_1 b), 3_1 b, 3_2 b])[1, 1, 1]).$$

For if $\mathbf{K}$, $\mathbf{L}$, $\mathbf{M}$ represent positive integers,

$$(\lambda b[S(3_1 b), 3_1 b, 3_2 b])[\mathbf{K}, \mathbf{L}, \mathbf{M}] \text{ conv } [S\mathbf{K}, \mathbf{K}, \mathbf{L}],$$

and hence if $\mathbf{A}$ represents a positive integer,

$$\mathbf{A}(\lambda b[S(3_1 b), 3_1 b, 3_2 b])[1, 1, 1] \text{ conv } [S\mathbf{A}, \mathbf{A}, \mathbf{B}],$$

where $\mathbf{B}$ represents the predecessor of the positive integer represented by $\mathbf{A}$. (The method of λ-definition of the predecessor function due to Kleene [35] is here modified by employment of a different formal representation of ordered triads.)

A kind of subtraction of positive integers, which we distinguish by placing a dot above the sign of subtraction, and which differs from the usual kind in that $x \dot{-} y = 1$ if $x \leqq y$, may now be shown to be λ-definable:

$$[\mathbf{M} \dot{-} \mathbf{N}] \rightarrow \mathbf{N}P\mathbf{M}.$$

The functions <u>the lesser of the two positive integers $x$ and $y$</u> and <u>the greater of the two positive integers $x$ and $y$</u> are λ-definable respectively by

$$\min \rightarrow \lambda ab \,.\, Sb \dot{-} \,.\, Sb \dot{-} a,$$

$$\max \rightarrow \lambda ab \,.\, [a+b] \dot{-} \min ab.$$

The parity of a positive integer, i.e., the function whose value is 1 for an odd positive integer and 2 for an even positive integer, is $\lambda$-defined by

$$\text{par} \rightarrow \lambda a.a(\lambda b.3 \dot{-} b)2.$$

Using ordered pairs in a way similar to that in which ordered triads were used to obtain a $\lambda$-definition of the predecessor function, we give a $\lambda$-definition of the function <u>the least integer not less than half of</u> $x$ -- or, in other words, the quotient upon dividing $x+1$ by 2, in the sense of division with a remainder:

$$H \rightarrow \lambda a.P(2_1(a(\lambda b[P[2_1 b + 2_2 b], 3 \dot{-} 2_2 b])[1, 2])).$$

Of course this $H$ is unrelated to the -- entirely different -- function $H$ which was introduced for illustration in §1.

If we let

$$\mathfrak{L} \rightarrow \lambda b.b(\lambda c\lambda d[dPc(\lambda e.e1I)(\lambda fg.fgS)c, dPc(\lambda h.h1IS)(\lambda ijk.kij(\lambda l.l1))d]),$$

$$\mathfrak{U} \rightarrow \lambda a.a\mathfrak{L}[1, 1],$$

$$Z \rightarrow \lambda a.2_2(\mathfrak{U}a),$$

$$Z' \rightarrow \lambda a.\mathfrak{U}a(\lambda bc.b \dot{-} c),$$

then, if $\boldsymbol{M}, \boldsymbol{N}$ represent the positive integers $m, n$ respectively, $\mathfrak{L}[\boldsymbol{M}, \boldsymbol{N}]$ conv $[S\boldsymbol{M}, 1]$ if $m \dot{-} n = 1$ and conv $[\boldsymbol{M}, S\boldsymbol{N}]$ if $m \dot{-} n > 1$; hence $\mathfrak{U}1, \mathfrak{U}2, \ldots$ are convertible respectively into

$$[2, 1], [3, 1], [3, 2], [4, 1], [4, 2], [4, 3], [5, 1], \ldots ;$$

hence $Z1, Z2, \ldots$ are convertible respectively into

$$1, 1, 2, 1, 2, 3, 1, 2, 3, 4, 1, 2, 3, 4, 5, \ldots ,$$

and $Z'1$, $Z'2$, ... are convertible respectively into

$$1,\ 2,\ 1,\ 3,\ 2,\ 1,\ 4,\ 3,\ 2,\ 1,\ 5,\ 4,\ 3,\ 2,\ 1,\ \ldots\ .$$

Thus the infinite sequence of ordered pairs,

$$[Z1,\ Z'1],\ [Z2,\ Z'2],\ [Z3,\ Z'3],\ \ldots,$$

contains all ordered pairs of positive integers, with no repetitions. The function whose value for the arguments $x, y$ is the number of the ordered pair $[x, y]$ in this enumeration is λ-defined by

$$\text{nr} \rightarrow \lambda ab\ .\ S(H[[a+b] \times P[a+b]]) \dot{-} b.$$

10. PROPOSITIONAL FUNCTIONS; THE KLEENE 𝔭-FUNCTION. By a propositional function we shall mean a function (of one or more variables) whose values are truth values — i.e., truth and falsehood. A property is a propositional function of one variable; a relation is a propositional function of two variables. The characteristic function associated with a propositional function is the function whose value is 2 when (i.e., for an argument or arguments for which) the value of the propositional function is truth, whose value is 1 when the value of the propositional function is falsehood, and which has no value otherwise.

A propositional function of positive integers will be said to be λ-definable if the associated characteristic function is a λ-definable function. (It can readily be shown that the choice of the particular integers 2 and 1 in the definition of characteristic function is here non-essential; the class of λ-definable propositional functions of positive integers remains unaltered if any other pair of distinct positive integers is substituted.)

In particular, the relations $>$ and $=$ between positive integers are λ-definable, as is shown by giving λ-definitions of the associated characteristic functions:

$$\text{exc} \rightarrow \lambda ab \,.\, \text{min}\ 2\ [Sa \dot{-} b].$$

$$\text{eq} \rightarrow \lambda ab \,.\, 4 \dot{-} \,.\, \text{exc}\ ab + \text{exc}\ ba.$$

From this follows the $\lambda$-definability of a great variety of properties and relations of positive integers which are expressible by means of equations and inequalities; conjunction, disjunction, and negation of equations and inequalities can be provided for by using min, max, and $\lambda a.3 \dot{-} a$ respectively.

We prove also the two following theorems from Kleene [35], and a third closely related theorem:

10 I. If $R$ is a $\lambda$-definable propositional function of $n+1$ positive integer arguments, then the function $F$ is $\lambda$-definable (1) whose value for the positive integer arguments $x_1, x_2, \ldots, x_n$ is the least positive integer $y$ such that $Rx_1x_2\ldots x_ny$ holds (i.e., has the value truth), provided that there is such a least positive integer $y$ and that, for every positive integer $z$ less than this $y$, $Rx_1x_2\ldots x_nz$ has a value, truth or falsehood, and (2) which has no value otherwise.

In the case that $R$ has a value for every set of $n+1$ positive integer arguments, $F$ may be described simply by saying that $Fx_1x_2\ldots x_n$ is the least positive integer $y$ such that $Rx_1x_2\ldots x_ny$ holds.

Let

$$\mathfrak{G} \rightarrow \lambda n.n(\lambda r.r(\lambda s.s\,1II(\lambda xgt.g1(tx)Ix)))$$
$$(\lambda f.fI1II)(\lambda xgt.g(t(Sx))(Sx)gt).$$

Then

$$\mathfrak{G}1 \text{ red } \lambda xgt.g(t(Sx))(Sx)gt,$$

$$\mathfrak{G}2 \text{ red } \lambda xgt.g1(tx)Ix.$$

Hence if $\boldsymbol{N}$ represents a positive integer and $\boldsymbol{TN}$ conv either 1 or 2, we have (using 7 XXVIII to show that $\boldsymbol{TN}$ red 1 or 2),

$$\mathfrak{G}1\boldsymbol{N}\mathfrak{G}\boldsymbol{T} \text{ red } \mathfrak{G}(\boldsymbol{T}(S\boldsymbol{N}))(S\boldsymbol{N})\mathfrak{G}\boldsymbol{T},$$

$$\mathfrak{G}2\boldsymbol{N}\mathfrak{G}\boldsymbol{T} \text{ red } \boldsymbol{N}.$$

Hence if we let

$$\mathfrak{p} \longrightarrow \lambda tx.\mathfrak{G}(tx)x\mathfrak{G}t,$$

we have $\mathfrak{p}\boldsymbol{TN}$ red $\boldsymbol{N}$ if $\boldsymbol{TN}$ conv 2, and $\mathfrak{p}\boldsymbol{TN}$ conv $\mathfrak{p}\boldsymbol{T}(S\boldsymbol{N})$ if $\boldsymbol{TN}$ conv 1, and (by 7 XXXI, 7 XXXII) $\mathfrak{p}\boldsymbol{TN}$ has no normal form if $\boldsymbol{TN}$ has no normal form.

If $\boldsymbol{N}$ represents the positive integer $n$ and $\boldsymbol{T}$ λ-defines the characteristic function associated with the property $T$ of positive integers, it follows that $\mathfrak{p}\boldsymbol{TN}$ is convertible into the formula which represents the least positive integer $y$, not less than $n$, for which $Ty$ holds, provided that there is such a least positive integer $y$ and that, for every positive integer $z$ less than this $y$ and not less than $n$, $Tz$ has a value, truth or falsehood; and that in any other case $\mathfrak{p}\boldsymbol{TN}$ has no normal form (in the case that $Ty$ has the value falsehood for all positive integers $y$ not less than $n$, we have

$$\mathfrak{p}\boldsymbol{TN} \text{ red } \mathfrak{G}(\boldsymbol{TN})\boldsymbol{N}\mathfrak{G}\boldsymbol{T} \text{ red } \mathfrak{G}(\boldsymbol{T}(S\boldsymbol{N}))(S\boldsymbol{N})\mathfrak{G}\boldsymbol{T} \text{ red } \mathfrak{G}(\boldsymbol{T}(S(S\boldsymbol{N})))(S(S\boldsymbol{N}))\mathfrak{G}\boldsymbol{T}$$
$$\text{red } \ldots$$

to infinity, and hence no normal form by 7 XXXI).

Let $\boldsymbol{R}$ be a formula which λ-defines the characteristic function associated with the propositional function $R$ referred to in 10 I. Then $F$ is λ-defined by

$$\lambda \boldsymbol{x}_1\boldsymbol{x}_2\ldots\boldsymbol{x}_n.\mathfrak{p}(\boldsymbol{R}\boldsymbol{x}_1\boldsymbol{x}_2\ldots\boldsymbol{x}_n)1.$$

10 II. If $T$ is a λ-definable property of positive integers, the function $F$ is λ-definable (1) whose value for the positive integer argument $x$ is the $x$th positive integer $y$ (in the order of magnitude of the positive integers) such that $Ty$ holds, provided that there is such a positive integer $y$ and that, for every positive integer $z$ less than $y$, $Tz$ has a value, truth or falsehood, and (2) which has no value otherwise.

For let $\mathbf{T}$ be a formula which $\lambda$-defines the characteristic function associated with $T$, and let

$$\mathfrak{P} \longrightarrow \lambda tx.P(x(\lambda n.S(\mathfrak{p}\, t\, n))1).$$

Then $\mathfrak{P}\mathbf{T}$ $\lambda$-defines $F$.

10 III. If $R_1$ and $R_2$ are $\lambda$-definable propositional functions each of n+1 positive integer arguments, then the propositional function $R$ is $\lambda$-definable

(1) whose value for the positive integer arguments $x_1$, $x_2$, ..., $x_n$ is falsehood if there is a positive integer $y$ such that $R_1x_1x_2\ldots x_ny$ holds and $R_1x_1x_2\ldots x_nz$ and $R_2x_1x_2\ldots x_nz$ both have the value falsehood for every positive integer $z$ less than $y$, and

(2) whose value for the positive integer arguments $x_1$, $x_2$. ..., $x_n$ is truth if there is a positive integer $y$ such that $R_2x_1x_2\ldots x_ny$ holds and $R_1x_1x_2\ldots x_ny$ has the value falsehood and $R_1x_1x_2\ldots x_nz$ and $R_2x_1x_2\ldots x_nz$ both have the value falsehood for every positive integer $z$ less than $y$, and

(3) which has no value otherwise.

Let

$$\text{alt} \longrightarrow \lambda xyn.\text{par}\,n(\lambda a.a(\lambda b.b1Iy))(\lambda c.c(\lambda def.fde))x(Hn).$$

$$\pi \longrightarrow \lambda xy.\text{par}(\mathfrak{p}(\text{alt}\ xy)1).$$

If $\mathbf{F}$ and $\mathbf{G}$ are functions of positive integers, each being a function of one argument and including the integer 1 in its range of arguments, and if $\mathbf{F}$ and $\mathbf{G}$ $\lambda$-define $F$ and $G$ respectively, then alt $\mathbf{FG}$ $\lambda$-defines the function whose value for the odd integer $2x-1$ is $\mathbf{F}x$ and whose value for the even integer $2x$ is $Gx$.

If $\mathbf{R}_1$ and $\mathbf{R}_2$ $\lambda$-define the characteristic functions associated with $R_1$ and $R_2$ respectively, then the characteristic function associated with $R$ is $\lambda$-defined by

$$\lambda x_1 x_2 \ldots x_n . \pi(R_1 x_1 x_2 \ldots x_n)(R_2 x_1 x_2 \ldots x_n)$$

-- this completes the proof of 10 III.

Formulas having the essential properties of $\mathfrak{p}$ and $\mathfrak{P}$ were first obtained by Kleene. These formulas λ-define (in a sense which will be readily understood without explicit definition) certain functions of functions of positive integers, as already indicated.

As a further application of the formula $\mathfrak{p}$, we give λ-definitions of subtraction of positive integers in the ordinary sense (so that $x-y$ has no value if $x \leq y$) and exact division (so that $x \div y$ has no value unless $x$ is a multiple of $y$):

$$[M-N] \rightarrow \mathfrak{p}(\lambda a . \text{ eq } M [N+a])1.$$

$$[M \div N] \rightarrow \mathfrak{p}(\lambda a . \text{ eq } M [N \times a])1.$$

11. DEFINITION BY RECURSION. A function $F$ of n positive integer arguments is said to be defined by composition in terms of the functions $G$ and $H_1, H_2, \ldots, H_m$ of positive integers (of the indicated numbers of arguments) by the equation,

$$Fx_1 x_2 \ldots x_n = G(H_1 x_1 x_2 \ldots x_n)(H_2 x_1 x_2 \ldots x_n) \ldots (H_m x_1 x_2 \ldots x_n).$$

(The case is not excluded that m or n or both are 1.)

A function $F$ of n+1 positive integer arguments is said to be defined by primitive recursion in terms of the functions $G_1$ and $G_2$ of positive integers (of the indicated numbers of arguments) by the pair of equations:

$$Fx_1 x_2 \ldots x_n 1 = G_1 x_1 x_2 \ldots x_n,$$

$$Fx_1 x_2 \ldots x_n (y+1) = G_2 x_1 x_2 \ldots x_n y (F x_1 x_2 \ldots x_n y).$$

(The case is not excluded that n = 0, the function $G_1$ being replaced in that case by a given positive integer $a$.)

The class of primitive recursive functions of positive integers is defined by the three following rules, a function being primitive recursive if and only if it is determined as such by these rules:

(1) The function $C$ such that $Cx = 1$ for every positive integer $x$, the successor function of positive integers, and the functions $U_i^n$ (where n is any positive integer and i is any positive integer not greater than n) such that $U_i^n x_1 x_2 \ldots x_n = x_i$, are primitive recursive.

(2) If the function $F$ of n arguments is defined by composition in terms of the functions $G$ and $H_1, H_2, \ldots, H_m$ and if $G, H_1, H_2, \ldots, H_m$ are primitive recursive, then $F$ is primitive recursive.

(3) If the function $F$ of n+1 arguments is defined by primitive recursion in terms of the functions $G_1$ and $G_2$ and if $G_1$ and $G_2$ are primitive recursive, then $F$ is primitive recursive; or in the case that n = 0, if $F$ is defined by primitive recursion in terms of the integer $a$ and the function $G_2$ and if $G_2$ is primitive recursive, then $F$ is primitive recursive.

In order to show that every primitive recursive function of positive integers is $\lambda$-definable, we must show that all the functions mentioned in (1) are $\lambda$-definable; that if $F$ is defined by composition in terms of $G$ and $H_1, H_2, \ldots, H_m$ and $G, H_1, H_2, \ldots, H_m$ are $\lambda$-definable, then $F$ is $\lambda$-definable; and that if $F$ is defined by primitive recursion in terms of $G_1$ and $G_2$ (or, in the case n = 0, in terms of $a$ and $G_2$) and if $G_1$ and $G_2$ are $\lambda$-definable (or, in the case n = 0, if $G_2$ is $\lambda$-definable), then $F$ is $\lambda$-definable.

Only the last of these three things makes any difficulty. Suppose that $F$ is defined by primitive recursion in terms of $G_1$ and $G_2$, and that $G_1$ and $G_2$ are $\lambda$-defined respectively by $\mathbf{G}_1$ and $\mathbf{G}_2$. Then in order to obtain a formula $\mathbf{F}$ which $\lambda$-defines $F$ we employ ordered triads:

$$\mathbf{F} \rightarrow \lambda \mathbf{x}_1 \mathbf{x}_2 \ldots \mathbf{x}_n \mathbf{y} . 3_3(\mathbf{y}(\lambda \mathbf{z}[S(3_1 \mathbf{z}), \mathbf{G}_2 \mathbf{x}_1 \mathbf{x}_2 \ldots \mathbf{x}_n (3_1 \mathbf{z})(3_2 \mathbf{z}), 3_2 \mathbf{z}])[1, \mathbf{G}_1 \mathbf{x}_1 \mathbf{x}_2 \ldots \mathbf{x}_n, 1])$$

($\mathbf{x}_1, \mathbf{x}_2, \ldots, \mathbf{x}_n, \mathbf{y}, \mathbf{z}$ being any n+2 distinct variables). In the case n = 0, this reduces to:

$$\mathbf{F} \rightarrow \lambda \mathbf{y}. \mathfrak{Z}_3(\mathbf{y}(\lambda \mathbf{z}[S(\mathfrak{Z}_1\mathbf{z}), \mathbf{G}_2(\mathfrak{Z}_1\mathbf{z})(\mathfrak{Z}_2\mathbf{z}), \mathfrak{Z}_2\mathbf{z}])[1, \mathbf{A}, 1]),$$

where $\mathbf{A}$ represents the positive integer $a$.

(The $\lambda$-definition of the predecessor function given in §9 may be regarded as a special case of the foregoing in which $a$ is 1 and $G_2$ is $U_1^2$. The extension of the method used for the predecessor function to the general case of definition by primitive recursion is due to Paul Bernays, in a letter of May 27th, 1935 -- where, however, the matter is stated within the context of the calculus of $\lambda$-$K$-conversion and ordered pairs are consequently used instead of ordered triads. As remarked by Bernays, this method of dealing with definition by primitive recursion has the advantage that it shows also, for each n, the $\lambda$-definability of the function $\rho$ of functions of positive integers whose value for the arguments $G_1$ and $G_2$ is the function $F$ defined by primitive recursion in terms of $G_1$ and $G_2$ -- i.e., essentially, the function $\rho$ of Hilbert [31].)

Thus we have:

11 I. Every primitive recursive function of positive integers is $\lambda$-definable.

The class of primitive recursive functions is known to include substantially all the ordinarily used numerical functions -- cf., e.g., Skolem [50], Gödel [27], Péter [41] (it is readily seen to be a non-essential difference that some of these authors deal with primitive recursive functions of non-negative integers rather than of positive integers). Primitive recursive, in particular, are functions corresponding to the quotient and remainder in division, the greatest common divisor, the $x$th prime number, and many related functions; $\lambda$-definitions of these functions can consequently be obtained by the method just given.

The two schemata, of definition by composition and by primitive recursion, have this property in common, that -- on the hypothesis that all particular values are known of the functions in terms of which $F$ is defined -- the given equations make possible the calculation of any required particular value of $F$ by a series of steps each consisting of a substitution, either of a (symbol for a) particular number for (all occurrences of) a variable, or of one thing for another known to be equal to it. By allowing additional, or more general, schemata having this property, various more extensive notions of recursiveness are obtainable (cf. Hilbert [31], Ackermann [1], Pēter [41, 42, 43, 44]). If the definition of primitive recursiveness is modified by allowing, in place of (2) and (3), any definition by a set of equations having this property, the functions obtained are called general recursive -- if it is required of all functions defined that they have a value for every set of the relevant number of positive integer arguments -- or partial recursive if this is not required. For a more exact statement (which may be made in any one of several equivalent ways), the reader is referred to Gödel [28], Church [9], Kleene [36, 39], Hilbert and Bernays [33].

That every general recursive function of positive integers is $\lambda$-definable can be proved in consequence of 10 I and 11 I by using the result of Kleene [36], that every general recursive function of n positive integer arguments $x_1, x_2, \ldots, x_n$ can be expressed in the form $F(\epsilon y(Rx_1 x_2 \ldots x_n y))$, where $F$ is a primitive recursive function of positive integers, $R$ is a propositional function of positive integers whose associated characteristic function is primitive recursive, and "$\epsilon y$" is to be read "the least positive integer $y$ such that." (Cf. Kleene [37]). The converse proposition, that every $\lambda$-definable function of positive integers, having a value for every set of the relevant number of positive integer arguments, is general recursive, is proved by the method of Church [9] or Kleene [37] (the proof makes use of the fact that, by 7 XXXI, the process of reduction to normal form provides a method of calculating explicitly any required particular value of a function whose $\lambda$-definition is given, and proceeds by setting up a set of recursion equations which in effect describe this process of calculation).

These proofs may be extended to the case of partial recursive functions without major modifications (cf. Kleene [39]). Hence are obtained the following theorems (proofs omitted here):

11 II. **Every partial recursive function of positive integers**

is λ-definable.

11 III. Every λ-definable function of positive integers is partial recursive.

The notion of a method of effective calculation of the values of a function, or the notion of a function for which such a method of calculation exists, is of not uncommon occurrence in connection with mathematical questions, but it is ordinarily left on the intuitive level, without attempt at explicit definition. The known theorems concerning λ-definability, or recursiveness, strongly suggest that the notion of an effectively calculable function of positive integers be given an exact definition by identifying it with that of a λ-definable function, or equivalently of a partial recursive function. As in all cases where a formal definition is offered of what was previously an intuitive or empirical idea, no complete proof is possible; but the writer has little doubt of the finality of the identification. (Concerning the origin of this proposal, see Church [9], footnotes 3, 18.)

An equivalent definition of effective calculability is to identify it with calculability within a formalized system of logic whose postulates and rules have appropriate properties of recursiveness -- cf. Church [9], §7, Hilbert and Bernays [33], Supplement II.

Another equivalent definition, having a more immediate intuitive appeal is that of Turing [55], who calls a function computable if (roughly speaking) it is possible to make a finite calculating machine capable of computing any required value of the function. The machine is supplied with a tape on which computations are printed (the analogue of the paper used by a human calculator), and no upper limit is placed on the length of tape or on the time required for computation of a particular value of the function, except that it be finite in each case. Further restrictions imposed on the character of the machine are more or less clearly either non-essential or necessarily contained in the requirement of finiteness. The equivalence of computability to λ-definability and general recursiveness (attention being confined to functions of one argument for which the range of arguments consists of all positive integers) is proved in Turing [57].

Mention should also be made of the notion of a finite combinatory process introduced by Post [46]. This again is equivalent to the other concepts of effective calculability.

Examples of functions which are not effectively calculable can now be given in various ways. In particular, it is proved in Church [9] that if the set of well-formed formulas of the calculus of λ-conversion be enumerated in a straightforward way (any one of the particular enumerations which immediately suggest themselves may be employed), and if $F$ is the function such that $F$ is 2 or 1 according as the $x$th formula in this enumeration has or has not a normal form, then $F$ is not λ-definable. This may be taken as the exact meaning of the somewhat vague statement made at the end of §6, that the condition of having a normal form is not effective.

In the explicit proofs of many of the theorems which have been stated without proof in this section, use is made of the notion of the Gödel number of a formula or formal expression. In the published papers referred to, this notion is introduced by a method closely similar to that employed by Gödel [27]. In the case of well-formed formulas of the calculus of λ-conversion, however, it would be equally possible to use the somewhat different method of our next chapter.

# Chapter IV

## COMBINATIONS, GÖDEL NUMBERS

12. COMBINATIONS. If s is any set of well-formed formulas, the class of <u>s-combinations</u> is defined by the two following rules, a formula being an s-combination if and only if it is determined as such by these rules:

(1) Any formula of the set s, and any variable standing alone, is an s-combination.

(2) If $\boldsymbol{A}$ and $\boldsymbol{B}$ are s-combinations, $\boldsymbol{AB}$ is an s-combination.

In the cases in which we shall be interested the formulas of s will contain no free variables and will none of them be of the form $\boldsymbol{AB}$. In such a case it is possible to distinguish the <u>terms</u> of an s-combination, each occurrence of a free variable or of one of the formulas of s being a term.

If s is the null set, the s-combinations will be called <u>combinations of variables</u>.

If s consists of the two formulas $I$, $J$, where

$$I \rightarrow \lambda a a,$$
$$J \rightarrow \lambda abcd.ab(adc),$$

the s-combinations will be called simply <u>combinations</u>.

We shall prove that <u>every well-formed formula is convertible into a combination</u>. This theorem is taken from Rosser [47], the present proof of it from Church [8]; the ideas involved go back to Schönfinkel [49] and Curry [18, 21].

Let:

$$\tau \rightarrow JII.$$

Then $\tau$ conv $\lambda ab.ba$, and hence $\tau \boldsymbol{AB}$ conv $\boldsymbol{BA}$.

If $M$ is any combination containing x as a free variable, we define an associated combination $\lambda_x M|$, which does not contain x as a free variable but otherwise contains the same free variables as $M$. This definition is by recursion, according to the following rules:

(1) $\lambda_x x|$ is $I$.
(2) If $B$ contains x as a free variable and $A$ does not, $\lambda_x AB|$ is $J\tau\lambda_x B|(JIA)$.
(3) If $A$ contains x as a free variable and $B$ does not, $\lambda_x AB|$ is $J\tau B\lambda_x A|$.
(4) If both $A$ and $B$ contain x as a free variable, $\lambda_x AB|$ is $J\tau\tau(JI(J\tau\tau(J\tau\lambda_x B|(J\tau\lambda_x A|J))))$.

12 I. If $M$ is a combination containing x as a free variable, $\lambda_x M|$ conv $\lambda x M$.

We prove this by induction with respect to the number of terms of $M$.

If $M$ has one term, then $M$ is x, and $\lambda_x M|$ is $I$, which is convertible into $\lambda xx$.

If $M$ is $AB$ and $B$ contains x as a free variable and $A$ does not, then $\lambda_x M|$ is $J\tau\lambda_x B|(JIA)$, which (see definitions of $I$, $J$, $\tau$) is convertible into $\lambda a.A(\lambda_x B|a)$, which, by hypothesis of induction, is convertible into $\lambda a.A((\lambda xB)a)$ which finally is convertible into $\lambda x.AB$.

If $M$ is $AB$ and $A$ contains x as a free variable and $B$ does not, then $\lambda_x M|$ is $J\tau B\lambda_x A|$, which is convertible into $\lambda a.\lambda_x A|aB$, which, by hypothesis of induction is convertible into $\lambda a.(\lambda xA)aB$, which finally is convertible into $\lambda x.AB$.

If $M$ is $AB$ and both $A$ and $B$ contain x as a free variable, then $\lambda_x M|$ is $J\tau\tau(JI(J\tau\tau(J\tau\lambda_x B|(J\tau\lambda_x A|J))))$, which is convertible into $\lambda a.\lambda_x A|a(\lambda_x B|a)$, which, by hypothesis of induction, is convertible into $\lambda a.(\lambda xA)a((\lambda xB)a)$, which finally is convertible into $\lambda x.AB$.

The foregoing tacitly assumes that $A$ and $B$ do not contain $a$ as a free variable. The modification necessary for the contrary case is, however, obvious.

This completes the proof of 12 I. We define the combination belonging to a well-formed formula, by recursion as follows:

(1) The combination belonging to $x$ is $x$ (where $x$ is any variable).

(2) The combination belonging to $FA$ is $F'A'$, where $F'$ and $A'$ are the combinations belonging to $F$ and $A$ respectively.

(3) The combination belonging to $\lambda x M$ is $\lambda_x M'|$, where $M'$ is the combination belonging to $M$.

12 II. Every well-formed formula is convertible into the combination belonging to it.

Using 12 I, this is proved by induction with respect to the length of the formula. The proof is straightforward and details are left to the reader.

12 III. The combination belonging to $X$ and the combination belonging to $Y$ are identical if and only if $X$ conv-I $Y$.

13. PRIMITIVE SETS OF FORMULAS. A set s of well-formed formulas is called a <u>primitive set</u>, if the formulas of s contain no free variables and are none of them of the form $AB$, and every well-formed formula is convertible into an s-combination. (When necessary to distinguish this idea from the analogous idea in the calculus of $\lambda$-$K$-conversion, the calculus of $\lambda$-$\delta$-conversion, etc. -- see Chapter V -- we may speak of primitive sets of $\lambda$-formulas, primitive sets of $\lambda$-$K$-formulas, primitive sets of $\lambda$-$\delta$-formulas, etc.)

It was proved in §12 that the formulas $I$, $J$ are a primitive set. Another primitive set of formulas, suggested by the work of Curry, consists of the four formulas $B$, $C$, $W$, $I$, where:

$$B \rightarrow \lambda abc.a(bc).$$

$$C \rightarrow \lambda abc.acb.$$

$$W \rightarrow \lambda ab.abb.$$

In order to prove this it is sufficient to express $J$ as a $\{B, C, W, I\}$-combination, as follows:

$$J \text{ conv } B(BC(BC))(B(W(BBB))C).$$

Still another primitive set of formulas consists of the four formulas $B$, $T$, $D$, $I$, where:

$$T \longrightarrow \lambda ab.ba.$$

$$D \longrightarrow \lambda a.aa.$$

In order to prove this it is sufficient to express $C$ and $W$ as $\{B, T, D, I\}$-combinations, as follows:

$$C \text{ conv } B(T(BBT))(BBT).$$

$$W \text{ conv } B(B(T(BD(B(TT)(B(BBB)T))))(BBT))(B(T(B(TI)(TI)))B).$$

A primitive set of formulas is said to be <u>independent</u> if it ceases to be a primitive set upon omission of any one of the formulas. It seems plausible that each of the three primitive sets which have been named is independent. -- In the case of the set $\{I, J\}$, the independence of $J$ follows (using 7 XVII) from the fact that any combination all of whose terms are $I$ is convertible into $I$; and the independence of $I$ follows (using 7 XXVIII) from the fact that if $\mathbf{A}$ imr $\mathbf{B}$ and $\mathbf{B}$ contains a (well-formed) part convertible-I into $I$ then $\mathbf{A}$ must contain a (well-formed) part convertible-I into $I$.

14. AN APPLICATION OF THE THEORY OF COMBINATIONS. We prove now the following theorems, due to Kleene [34, 35, 37]:

14 I. If $\mathbf{A}_1$ and $\mathbf{A}_2$ contain no free variables, a formula $\mathbf{L}$ can be found such that $\mathbf{L}1$ conv $\mathbf{A}_1$ and $\mathbf{L}2$ conv $\mathbf{A}_2$.

For, by 12 II, $\mathbf{A}_1$ and $\mathbf{A}_2$ are convertible into combinations $\mathbf{A}_1'$ and $\mathbf{A}_2'$ respectively. We take $\mathbf{A}_1'$ to be the combination belonging to $\mathbf{A}_1$, unless that combination fails to contain an occurrence of $J$, in which case we take $\mathbf{A}_1'$ to be $JIIII$; and $\mathbf{A}_2'$ is similarly determined relatively to $\mathbf{A}_2$. Let $\mathbf{A}_1''$ and $\mathbf{A}_2''$ be the result of replacing all occurrences of $J$ by the variable $j$ in $\mathbf{A}_1'$ and $\mathbf{A}_2'$ respectively, and let $\mathbf{B}_1$ and

$\mathbf{B}_2$ be $\lambda j \mathbf{A}_1''$ and $\lambda j \mathbf{A}_2''$ respectively. Then $\mathbf{B}_1 J$ conv $\mathbf{A}_1$, and $\mathbf{B}_2 J$ conv $\mathbf{A}_2$, and $\mathbf{B}_1 I$ conv $I$, and $\mathbf{B}_2 I$ conv $I$. Consequently a formula $\mathbf{L}$ having the required property is:

$$\lambda n.n(\lambda x.x(\lambda y.yI\mathbf{B}_2))(\lambda z.zII)\mathbf{B}_1 J.$$

14 II. If $\mathbf{A}_1, \mathbf{A}_2, \ldots, \mathbf{A}_n$ contain no free variables, a formula $\mathbf{L}$ can be found such that $\mathbf{L}1$ conv $\mathbf{A}_1$, $\mathbf{L}2$ conv $\mathbf{A}_2$, ..., $\mathbf{L}\mathbf{N}$ conv $\mathbf{A}_n$($\mathbf{N}$ being the formula which represents n ).

For the case that n is 1 or 2, this follows from 14 I. For larger values of n, we prove it by induction.

Let $\mathbf{L}_2$ be a formula such that $\mathbf{L}_2 1$ conv $\mathbf{A}_1$, and let $\mathbf{L}_1$ be a formula such that $\mathbf{L}_1 1$ conv $\mathbf{A}_2$, $\mathbf{L}_1 2$ conv $\mathbf{A}_3$, ..., $\mathbf{L}_1\mathbf{M}$ conv $\mathbf{A}_n$ (where $\mathbf{M}$ represents n-1). Also let $\mathbf{C}$ be a formula such that $\mathbf{C}1$ conv $\mathbf{L}_1$ and $\mathbf{C}2$ conv $\mathbf{L}_2$. Then a formula $\mathbf{L}$ having the required property is:

$$\lambda i.\mathbf{C}[\mathfrak{3} \dot{-} i](\mathbf{P}i).$$

14 III. If $\mathbf{A}_1, \mathbf{A}_2, \ldots, \mathbf{A}_n, \mathbf{F}_1, \mathbf{F}_2, \ldots, \mathbf{F}_m$ contain no free variables, a formula $\mathbf{E}$ can be found which represents an enumeration of the least set of formulas which contains $\mathbf{A}_1, \mathbf{A}_2, \ldots, \mathbf{A}_n$ and is closed under each of the operations of forming $\mathbf{F}_\alpha \mathbf{X}\mathbf{Y}$ from the formulas $\mathbf{X},\mathbf{Y}$ ($\alpha$ = 1, 2, ..., n ), in the sense that every formula of this set is convertible into one of the formulas in the infinite sequence

$$\mathbf{E}1,\ \mathbf{E}2,\ \ldots,$$

and every formula in this infinite sequence is convertible into one of the formulas of the set.

We prove this first for the case m = 1, using a device due to Kleene for obtaining formulas satisfying arbitrary conversion conditions of the general kind illustrated in (1) below.

Using 14 II, let $\mathbf{U}$ be a formula such that

$$\boldsymbol{U}1 \text{ conv } I,$$

$$\boldsymbol{U}2 \text{ conv } \lambda xy.\boldsymbol{F}_1(y(S[\boldsymbol{N}' \dot{-} Zx])[Zx \dot{-} \boldsymbol{N}]y)(y(S[\boldsymbol{N}' \dot{-} Z'x])[Z'x \dot{-} \boldsymbol{N}]y),$$

$$\boldsymbol{U}3 \text{ conv } \lambda xy.yx\boldsymbol{A}_1,$$

$$\boldsymbol{U}4 \text{ conv } \lambda xy.yx\boldsymbol{A}_2,$$

$$\cdots\cdots\cdots\cdots\cdots\cdots$$

$$\boldsymbol{UN}' \text{ conv } \lambda xy.yx\boldsymbol{A}_n,$$

where $\boldsymbol{N}$ represents n and $\boldsymbol{N}'$ represents n+2, and $Z$ and $Z'$ are the formulas introduced in §9. Let $\boldsymbol{E}$ be the formula,

$$\lambda i.\boldsymbol{U}(S[\boldsymbol{N}' \dot{-} i])[i \dot{-} \boldsymbol{N}]\boldsymbol{U}.$$

Then we have:

$$\boldsymbol{E}1 \text{ conv } \boldsymbol{A}_n,$$

$$\boldsymbol{E}2 \text{ conv } \boldsymbol{A}_{n-1},$$

(1) $$\cdots\cdots\cdots\cdots\cdots\cdots$$

$$\boldsymbol{EN} \text{ conv } \boldsymbol{A}_1,$$

$$\boldsymbol{EK} \text{ conv } \boldsymbol{F}_1(\boldsymbol{E}(Z[\boldsymbol{K} \dot{-} \boldsymbol{N}]))(\boldsymbol{E}(Z'[\boldsymbol{K} \dot{-} \boldsymbol{N}])),$$

$\boldsymbol{K}$ being any formula which represents an integer greater than n. From this it follows that $\boldsymbol{E}$ is a formula of the kind required.

Consider now the case $m > 1$. Let $\boldsymbol{M}$ represent m and let $\boldsymbol{F}$ be a formula such that $\boldsymbol{F}1$ conv $\boldsymbol{F}_1$, $\boldsymbol{F}2$ conv $\boldsymbol{F}_2$, ..., $\boldsymbol{FM}$ conv $\boldsymbol{F}_m$. By the preceding proof for the case $m = 1$, a formula $\boldsymbol{E}'$ can be found which represents an enumeration of the least set of formulas which contains $[1, \boldsymbol{A}_1]$, $[2, \boldsymbol{A}_1]$, ..., $[\boldsymbol{M}, \boldsymbol{A}_1]$, $[1, \boldsymbol{A}_2]$, $[2, \boldsymbol{A}_2]$, ..., $[\boldsymbol{M}, \boldsymbol{A}_2]$, ..., $[1, \boldsymbol{A}_n]$, $[2, \boldsymbol{A}_n]$, ..., $[\boldsymbol{M}, \boldsymbol{A}_n]$ and is closed under the operation of forming $\boldsymbol{V}(\lambda xy[x, \boldsymbol{XF}y])$ from the formulas $\boldsymbol{X}, \boldsymbol{Y}$. Then a formula $\boldsymbol{E}$ of the kind required is:

$$\lambda i.2_2(\boldsymbol{E}'i).$$

It is immaterial that the enumeration so obtained contains repetitions. (Notice that $2_2[\boldsymbol{B}, \boldsymbol{C}]$ conv $\boldsymbol{C}$ if $\boldsymbol{B}$ is any formula such that $\boldsymbol{B}I$ conv $I$, in particular if $\boldsymbol{B}$ is any formula representing a positive integer; the case considered in §9 that $\mathbf{B}$ and $\mathbf{C}$ <u>both</u> represent positive integers is thus only a special case.)

14 IV. If $\mathbf{A}_1, \mathbf{A}_2, \ldots, \mathbf{A}_n, \mathbf{F}_1, \mathbf{F}_2, \ldots, \mathbf{F}_m, \mathbf{F}_{m+1}, \mathbf{F}_{m+2}, \ldots, \mathbf{F}_{m+r}$ contain no free variables, a formula $\mathbf{E}$ can be found which represents an enumeration of the least set of formulas which contains $\mathbf{A}_1, \mathbf{A}_2, \ldots, \mathbf{A}_n$ and is closed under each of the operations of forming $\mathbf{F}_\alpha \mathbf{X}\mathbf{Y}$ from the formulas $\mathbf{X}, \mathbf{Y}$ ($\alpha = 1, 2, \ldots, m$) and of forming $\mathbf{F}_{m+\beta}\mathbf{X}$ from the formula $\mathbf{X}$ ($\beta = 1, 2, \ldots, r$) -- in the sense that every formula of this set is convertible into one of the formulas in the infinite sequence

$$\mathbf{E}1, \mathbf{E}2, \ldots,$$

and every formula in this infinite sequence is convertible into one of the formulas of the set.

(The case is not excluded that $m = 0$ or that $r = 0$, provided that $m$ and $r$ are not both $0$.)

By the method used in the proof of 14 I, find formulas $\mathbf{B}_1, \mathbf{B}_2, \ldots, \mathbf{B}_n, \mathbf{C}_1, \mathbf{C}_2, \ldots, \mathbf{C}_{m+r}$ such that $\mathbf{B}_1 J$ conv $\mathbf{A}_1$, $\mathbf{B}_2 J$ conv $\mathbf{A}_2$, $\ldots$, $\mathbf{B}_n J$ conv $\mathbf{A}_n$, $\mathbf{C}_1 J$ conv $\mathbf{F}_1$, $\mathbf{C}_2 J$ conv $\mathbf{F}_2$, $\ldots$, $\mathbf{C}_{m+r} J$ conv $\mathbf{F}_{m+r}$, and $\mathbf{B}_1 I$ conv $I$, $\mathbf{B}_2 I$ conv $I$, $\ldots$, $\mathbf{B}_n I$ conv $I$, $\mathbf{C}_1 I$ conv $I$, $\mathbf{C}_2 I$ conv $I$, $\ldots$, $\mathbf{C}_{m+r} I$ conv $I$. By 14 III, a formula $\mathbf{E}'$ can be found which represents an enumeration of the least set of formulas which contains $\mathbf{B}_1, \mathbf{B}_2, \ldots, \mathbf{B}_n$ and is closed under each of the operations of forming $\lambda x.\mathbf{C}_\alpha x(\mathbf{X}x)(\mathbf{Y}x)$ from the formulas $\mathbf{X}, \mathbf{Y}$ ($\alpha = 1, 2, \ldots, m$) and of forming $\lambda x.\mathbf{Y}I\mathbf{C}_{m+\beta}x(\mathbf{X}x)$ from the formulas $\mathbf{X}, \mathbf{Y}$ ($\beta = 1, 2, \ldots, r$). Then a formula $\mathbf{E}$ of the kind required is:

$$\lambda i.\mathbf{E}' i J.$$

15. A COMBINATORY EQUIVALENT OF CONVERSION. It is desirable to have a set of operations (upon combinations) which have the property that they always change a combination into a combination and which constitute an equivalent of conversion in the sense that a combination $\mathbf{X}$ can be changed into a combination

$\mathbf{Y}$ by a sequence of (0 or more of) these operations if and only if $\mathbf{X}$ conv $\mathbf{Y}$. Such a set of operations is the following (OI - OXXXVIII) -- where $\mathbf{F}, \mathbf{A}, \mathbf{B}, \mathbf{C}, \mathbf{D}$ are arbitrary combinations, $\beta$, $\gamma$, $\omega$ are defined as indicated below, and the sign $\vdash$ is used to mean that the combination which precedes $\vdash$ is changed by the operation into the combination which follows:

OI. $I\mathbf{A} \vdash \mathbf{A}$.

OII. $\mathbf{A} \vdash I\mathbf{A}$.

OIII. $\mathbf{F}(I\mathbf{A}) \vdash \mathbf{FA}$.

OIV. $\mathbf{FA} \vdash \mathbf{F}(I\mathbf{A})$.

OV. $\mathbf{F}(I\mathbf{AB}) \vdash \mathbf{F}(\mathbf{AB})$.

OVI. $\mathbf{F}(\mathbf{AB}) \vdash \mathbf{F}(I\mathbf{AB})$.

OVII. $\mathbf{F}(J\mathbf{ABCD}) \vdash \mathbf{F}(\mathbf{AB}(\mathbf{ADC}))$.

OVIII. $\mathbf{F}(\mathbf{AB}(\mathbf{ADC})) \vdash \mathbf{F}(J\mathbf{ABCD})$.

OIX. $\mathbf{F}J \vdash \mathbf{F}(\omega(\beta\gamma(\beta(\beta(\beta\gamma))(\beta(\beta(\beta\beta\beta))I)))))$.

OX. $\mathbf{F}(\omega(\beta\gamma(\beta(\beta(\beta\gamma))(\beta(\beta(\beta\beta\beta))I))))) \vdash \mathbf{F}J$.

OXI. $\mathbf{F}\beta \vdash \mathbf{F}(\beta(\beta(\beta I))\beta)$.

OXII. $\mathbf{F}(\beta(\beta(\beta I))\beta) \vdash \mathbf{F}\beta$.

OXIII. $\mathbf{F}\gamma \vdash \mathbf{F}(\beta(\beta(\beta I))\gamma)$.

OXIV. $\mathbf{F}(\beta(\beta(\beta I))\gamma) \vdash \mathbf{F}\gamma$.

OXV. $\mathbf{F}I \vdash \mathbf{F}(\beta II)$.

OXVI. $\mathbf{F}(\beta II) \vdash \mathbf{F}I$.

OXVII. $\mathbf{F}(\gamma(\beta\beta(\beta\beta\beta))\beta) \vdash \mathbf{F}(\beta(\beta\beta)\beta)$.

OXVIII. $\mathbf{F}(\beta(\beta\beta)\beta) \vdash \mathbf{F}(\gamma(\beta\beta(\beta\beta\beta))\beta)$.

OXIX. $\mathbf{F}(\gamma(\beta\beta(\beta\beta\beta))\gamma) \vdash \mathbf{F}(\beta(\beta\gamma)(\beta\beta\beta))$.

OXX. $\mathbf{F}(\beta(\beta\gamma)(\beta\beta\beta)) \vdash \mathbf{F}(\gamma(\beta\beta(\beta\beta\beta))\gamma)$.

OXXI. $\mathbf{F}(\gamma(\beta\beta\beta)\omega) \vdash \mathbf{F}(\beta(\beta\omega)(\beta\beta\beta))$.

OXXII. $\mathbf{F}(\beta(\beta\omega)(\beta\beta\beta)) \vdash \mathbf{F}(\gamma(\beta\beta\beta)\omega)$.

OXXIII. $\mathbf{F}(\gamma\beta I) \vdash \mathbf{F}(\beta(\beta I)I)$.

OXXIV. $\mathbf{F}(\beta(\beta I)I) \vdash \mathbf{F}(\gamma\beta I)$.

OXXV. $\mathbf{F}(\beta\beta\gamma) \vdash \mathbf{F}(\beta(\beta(\beta\gamma)\gamma)(\beta\beta))$.

OXXVI. $\mathbf{F}(\beta(\beta(\beta\gamma)\gamma)(\beta\beta)) \vdash \mathbf{F}(\beta\beta\gamma)$.

OXXVII. $\mathbf{F}(\beta\beta\omega) \vdash \mathbf{F}(\beta(\beta(\beta(\beta(\beta\omega)\omega)(\beta\gamma))(\beta(\beta\beta)))\beta)$.

OXXVIII. $\mathbf{F}(\beta(\beta(\beta(\beta(\beta\omega)\omega)(\beta\gamma))(\beta(\beta\beta)))\beta) \vdash \mathbf{F}(\beta\beta\omega)$.

OXXIX. $\mathbf{F}(\beta\gamma\gamma) \vdash \mathbf{F}(\beta(\beta I))$.

OXXX. $\mathbf{F}(\beta(\beta I)) \vdash \mathbf{F}(\beta\gamma\gamma)$.

OXXXI. $\mathbf{F}(\beta(\beta(\beta\gamma)\gamma)(\beta\gamma)) \vdash \mathbf{F}(\beta(\beta\gamma(\beta\gamma))\gamma)$.

OXXXII. $\mathbf{F}(\beta(\beta\gamma(\beta\gamma))\gamma) \vdash \mathbf{F}(\beta(\beta(\beta\gamma)\gamma)(\beta\gamma))$.

OXXXIII. $F(\beta\gamma\omega) \vdash F(\beta(\beta(\beta\omega)\gamma)(\beta\gamma))$.
OXXXIV. $F(\beta(\beta(\beta\omega)\gamma)(\beta\gamma)) \vdash F(\beta\gamma\omega)$.
OXXXV. $F(\beta\omega\gamma) \vdash F\omega$.
OXXXVI. $F\omega \vdash F(\beta\omega\gamma)$.
OXXXVII. $F(\beta\omega\omega) \vdash F(\beta\omega(\beta\omega))$.
OXXXVIII. $F(\beta\omega(\beta\omega)) \vdash F(\beta\omega\omega)$.

$$\gamma \longrightarrow J\tau(J\tau)(J\tau).$$
$$\beta \longrightarrow \gamma(JI\gamma)(JI).$$
$$\omega \longrightarrow \gamma(\gamma(\beta\gamma(\gamma(\beta J\tau)\tau))\tau).$$

(Note that $\tau$, $\gamma$, $\beta$, $\omega$ are convertible respectively into $T$, $C$, $B$, $W$.)

These thirty-eight operations have characteristics of simplicity not possessed by the operations I, II, III of §6, namely: (1) they are one-valued, i.e., given the combination operated on and the particular one of the thirty-eight operations which is applied, the combination resulting is uniquely determined; (2) they do not involve the idea of substitution at an arbitrary place, but only that of substitution at a specified place. This has the effect of rendering some of the developments in §16 much simpler than they otherwise might be.

The proof of the equivalence of OI-OXXXVIII to conversion is too long to be included here. It may be found in Rosser's dissertation [47] (cf. Section H therein). Many of the important ideas and methods involved derive from Curry [17, 18, 20, 21]; in fact, Curry has results which may be thought of as constituting an approximate equivalent to the one in question here but which are nevertheless sufficiently different so that we are unable to use them directly.

16. GÖDEL NUMBERS. The <u>Gödel number of</u> a combination is defined by induction as follows:

(1) The Gödel number of $I$ is 1.
(2) The Gödel number of $J$ is 3.
(3) The Gödel number of the $n$th variable in alphabetical order (see §5) is $2n+5$.
(4) If $m$ and $n$ are the Gödel numbers of $A$ and $B$ respec-

tively, the Gödel number of $AB$ is $(m+n)(m+n-1)-2n+2$.

The Gödel number belonging to a formula is defined to be the Gödel number of the combination belonging to the formula. (Notice that the Gödel number belonging to a combination is thus in general not the same as the Gödel number of the combination.)

It is left to the reader to verify that the Gödel numbers of two combinations $A$ and $B$ are the same if and only if $A$ and $B$ are the same; and that the Gödel numbers belonging to two formulas $A$ and $B$ are the same if and only if $A$ conv-I $B$ (cf. 12 III). (Notice that the Gödel number of $AB$, according to (4), is twice the number of the ordered pair $[m, n]$ in the enumeration of ordered pairs described at the end of §9.)

The usefulness of Gödel numbers arises from the fact that our formalism contains no notations for formulas -- i.e., for sequences of symbols. (It is not possible to use formulas as notations for themselves, because interconvertible formulas must denote the same thing although they are not the same formula, and because formulas containing free variables cannot denote any [fixed] thing.) The Gödel number belonging to a formula serves in many situations as a substitute for a notation for the formula and often enables us to accomplish things which might have been thought to be impossible without a formal notation for formulas.

This use of Gödel numbers is facilitated by the existence of a formula, form, such that, if $N$ represents the Gödel number belonging to $A$, and $A$ contains no free variables, then, form $N$ conv $A$. In order to obtain this formula, first notice that par $N$ conv 2 if $N$ represents the Gödel number of a combination having more than one term, and par $N$ conv 1 if $N$ represents the Gödel number of a combination having only one term; also that if $N$ represents the Gödel number of a combination $AB$, then $Z(HN)$ is convertible into the formula representing the Gödel number of $A$, and $Z'(HN)$ is convertible into the formula representing the Gödel number of $B$ (see §9). We introduce the abbreviations:

$$N_1 \rightarrow Z(HN).$$

$$N_2 \rightarrow Z'(HN).$$

Subscripts used in this way may be iterated, so that, for instance,

$$N_{122} \rightarrow Z'(H(Z'(H(Z(HN))))).$$

By the method of §14, find a formula $\mathfrak{D}$ such that

$$\mathfrak{D}1 \text{ conv } \lambda x.x12.$$

$$\mathfrak{D}2 \text{ conv } I,$$

$$\mathfrak{D}3 \text{ conv } \lambda x.x12J,$$

and a formula $\mathfrak{U}$ such that

$$\mathfrak{U}1 \text{ conv } \mathfrak{D},$$

$$\mathfrak{U}2 \text{ conv } \lambda xy.y(\text{par } x_1)x_1y(y(\text{par } x_2)x_2y),$$

(these formulas $\mathfrak{D}$ and $\mathfrak{U}$ can be explicitly written down by referring to the proofs of 14 I and 14 II).

Let

$$\text{form} \rightarrow \lambda n.\mathfrak{U}(\text{par } n)n\mathfrak{U}.$$

Then

$$\text{form } 1 \text{ conv } I,$$

$$\text{form } 3 \text{ conv } J, \text{ and}$$

$$\text{form } \boldsymbol{N} \text{ conv form } \boldsymbol{N}_1(\text{form } \boldsymbol{N}_2)$$

if $\boldsymbol{N}$ represents an even positive integer. From this it follows that form has the property ascribed to it above; for if $\boldsymbol{N}$ represents the Gödel number of a combination $\boldsymbol{A}'$ belonging to a formula $\boldsymbol{A}$, containing no free variables, then form $\boldsymbol{N}$ conv $\boldsymbol{A}'$, and $\boldsymbol{A}'$ conv $\boldsymbol{A}$.

Let:

$$\begin{aligned}\sigma \rightarrow \lambda n\,.\ & [\text{par}n + \text{par}n_1 + \text{eq}\overline{24812}n_{11} + [3 \dot{-} \text{eq}\overline{156}n_{12}] + \text{par}n_2 + \text{eq}\overline{12}n_{21} \dot{-} \overline{10}] \\ & + [2 \times [\text{par}n + \text{par}n_1 + \text{eq}\overline{24812}n_{11} + [3 \dot{-} \text{min}(\text{par}n_2)(\text{eq}\overline{12}n_{21})] \dot{-} 6]] \\ & + [3 \times [\text{par}n + \text{eq}\overline{623375746}n_1 + \text{par}n_2 + \text{eq}\overline{12}\,n_{21} + \text{par}n_{22} \\ & \qquad + \text{eq}\overline{623375746}n_{221} + \text{par}n_{222} + \text{par}n_{2221} + \text{eq}\overline{24812}n_{22211} \\ & \qquad + \text{par}n_{2222} + \text{par}n_{22221} + \text{eq}\overline{24812}n_{222211} + \text{eq}3n_{22222} \dot{-} \overline{24}]] \\ & \dot{-}\ 5.\end{aligned}$$

Noting that the Gödel numbers of $JI$, $\tau$, $J\tau$, $J\tau\tau$ are respectively 12, 156, 24812, 623375746, the reader may verify that:

$\sigma\boldsymbol{N}$ conv 1, 2, 3, or 4 if $\boldsymbol{N}$ represents a positive integer;

$\sigma N$ conv 2 if $N$ represents the Gödel number of a combination of the form $J\tau B(JIA)$, with $B$ different from $\tau$;

$\sigma N$ conv 3 if $N$ represents the Gödel number of a combination of the form $J\tau BA$ but not of the form $J\tau B(JIA)$;

$\sigma N$ conv 4 if $N$ represents the Gödel number of a combination of the form $J\tau\tau(JI(J\tau\tau(J\tau B(J\tau AJ))))$;

$\sigma N$ conv 1 if $N$ represents the Gödel number of a combination not of one of these three forms.

Again using §14, we find a formula u such that

u1 conv $\lambda xy.y5x$,
u2 conv $\lambda xy.y(\sigma x_{12})x_{12}y$,
u3 conv $\lambda xy.y(\sigma x_2)x_2y$,
u4 conv $\lambda xy.\min(y(\sigma x_{22212})x_{22212}y)(y(\sigma x_{222212})x_{222212}y)$,
u5 conv $\lambda x.3 \dot{-} x$,

and we let

$$o \rightarrow \lambda n.u(\sigma n)nu.$$

Then o $\lambda$-defines a function of positive integers whose value is 2 for an argument which is the Gödel number of a combination of the form $\lambda_x M|$, and 1 for an argument which is the Gödel number of a combination not of this form -- or, as we shall say briefly, o $\lambda$-defines the property of a combination of being of the form $\lambda_x M|$.

By similar constructions, involving lengthy detail but nothing new in principle, the following formulas may be obtained:

1) A formula, occ; such that, if $N$ represents a positive integer $n$, we have that occ $N$ $\lambda$-defines the property of a combination of containing the $n$th variable in alphabetical order, as a free variable (i.e., as a term).

2) A formula $\epsilon$, such that, $N$ representing a positive integer $n$, if $C$ represents the Gödel number of a combination not of the form $\lambda_x M|$, then $\epsilon NC$ conv $C$, and if $C$ represents the Gödel number of a combination $\lambda_x M|$, then $\epsilon NC$ is convertible into the formula representing the Gödel number of the combination obtained from $M$ by substituting for all free occurrences of x in $M$ the $n$th variable in alphabetical order.

3) A formula $\mathfrak{C}$, such that, if $C$ represents the Gödel

number of a combination not of the form $\lambda_x \boldsymbol{M}|$, then $\mathfrak{C}\boldsymbol{C}$ conv $\boldsymbol{C}$, and if $\boldsymbol{C}$ represents the Gödel number of a combination $\lambda_x \boldsymbol{M}|$, then $\mathfrak{C}\boldsymbol{C}$ is convertible into the formula representing the Gödel number of the combination obtained from $\boldsymbol{M}$ by substituting for all free occurrences of $\boldsymbol{x}$ in $\boldsymbol{M}$ the first variable in alphabetical order which does not occur in $\boldsymbol{M}$ as a free variable.

4) A formula r which λ-defines the property of a combination, that there is a formula to which it belongs.

5) A formula Λ which λ-defines the property of a combination of belonging to a formula of the form $\lambda \boldsymbol{x}\boldsymbol{M}$.

6) A formula, prim, which λ-defines the property of a combination of containing no free variables.

7) A formula, norm, which λ-defines the property of a combination of belonging to a formula which is in normal form.

8) A formula $0_1$ which corresponds to the operation OI of §15, in the sense that, if $\boldsymbol{C}$ represents the Gödel number of a combination of such a form that OI is not applicable to it, then $0_1\boldsymbol{C}$ conv $\boldsymbol{C}$, and if $\boldsymbol{C}$ represents the Gödel number of a combination $\boldsymbol{M}$ to which OI is applicable, then $0_1\boldsymbol{C}$ is convertible into the formula representing the Gödel number of the combination obtained from $\boldsymbol{M}$ by applying OI.

9) Formulas $0_2$, $0_3$, ..., $0_{38}$ which correspond respectively to the operations OII, OIII, ..., OXXXVIII of §15, in the same sense.

By 14 III, a formula, cb, can be found which represents an enumeration of the least set of formulas which contains 1 and 3 and is closed under the operation of forming $(\lambda ab \,.\, 2 \times \text{nr}\, ab)\boldsymbol{X}\boldsymbol{Y}$ from the formulas $\boldsymbol{X}$, $\boldsymbol{Y}$. But if $\boldsymbol{X}$, $\boldsymbol{Y}$ represent the Gödel numbers of combinations $\boldsymbol{A}$, $\boldsymbol{B}$ respectively, then $(\lambda ab \,.\, 2 \times \text{nr}\, ab)\boldsymbol{X}\boldsymbol{Y}$ is convertible into the formula which represents the Gödel number of $\boldsymbol{AB}$. Hence the formula, cb, enumerates the Gödel numbers of combinations containing no free variables, in the sense that every formula representing such a Gödel number is convertible into one of the formulas in the infinite sequence

$$\text{cb } 1, \text{ cb } 2, \ldots,$$

and every formula in this infinite sequence is convertible into a formula representing such a Gödel number.

If now we let

$$\text{ncb} \rightarrow \lambda n \,.\, \text{cb}\, (\mathfrak{P}(\lambda x \,.\, \text{norm}\, (\text{cb}\, x))n),$$

then ncb enumerates, in the same sense, the Gödel numbers of combinations which belong to formulas in normal form and contain no free variables (cf. 10 II).

By 14 IV, a formula 0 can be found which represents an enumeration of the least set of formulas which contains $I$ and is closed under each of the thirty-eight operations of forming $(\lambda ab.0_\beta(ab))\boldsymbol{X}$ from the formula $\boldsymbol{X}$ ($\beta = 1, 2, \ldots, 38$). Let

$$\text{cnvt} \rightarrow \lambda ab.0ba.$$

Then if $\boldsymbol{C}$ represents the Gödel number of a combination $\boldsymbol{M}$, the formula, cnvt $\boldsymbol{C}$, enumerates (again in the same sense as in the two preceding paragraphs) the Gödel numbers of combinations obtainable from $\boldsymbol{M}$ by conversion -- cf. §15.

Let

$$\text{nf} \rightarrow \lambda n \,.\, \text{cnvt}\; n(\mathfrak{p}(\lambda x \,.\, \text{norm}\, (\text{cnvt}\; nx))1).$$

Then nf λ-defines the operation <u>normal form of</u> a formula, in the sense that (1) if $\boldsymbol{C}$ represents the Gödel number of a combination $\boldsymbol{M}$, then nf $\boldsymbol{C}$ is convertible into the formula representing the Gödel number belonging to the normal form of $\boldsymbol{M}$; and hence (2) if $\boldsymbol{C}$ represents the Gödel number belonging to a formula $\boldsymbol{M}$, then nf $\boldsymbol{C}$ is convertible into the formula representing the Gödel number belonging to the normal form of $\boldsymbol{M}$. If $\boldsymbol{C}$ represents the Gödel number of a combination (or belonging to a formula) which has no normal form, then nf $\boldsymbol{C}$ has no normal form (cf. 10 I).

Let $\mathfrak{i}$ and $\mathfrak{s}$ be the formulas representing the Gödel numbers belonging 1 and $S$ respectively. Then the formulas

$$Z'(H(1(\lambda x \,.\, 2 \times \text{nr}\,\mathfrak{s}x)\mathfrak{i})),\; Z'(H(2(\lambda x \,.\, 2 \times \text{nr}\;\mathfrak{s}x)\mathfrak{i})),$$
$$Z'(H(3(\lambda x \,.\, 2 \times \text{nr}\;\mathfrak{s}x)\mathfrak{i})), \ldots,$$

are convertible respectively into formulas representing Gödel numbers belonging to

$$1,\; S1,\; S(S1),\; \ldots\;.$$

Hence a formula $\nu$ which λ-defines the property of a combination of belonging to a formula in normal form which represents a positive integer, may be obtained by defining:

$$\nu \rightarrow \lambda n\ .\ \pi(\mathrm{eq}\ n)(\lambda m\ .\ \mathrm{eq}\ n(\mathrm{nf}(\boldsymbol{Z}'(\boldsymbol{H}(m(\lambda x\ .\ 2 \times \mathrm{nr}\ {}_{\mathfrak{s}}x)\mathfrak{i}))))).$$

(It is necessary, in order to see this, to refer to 10 III, and to observe that the Gödel number belonging to a formula in normal form representing a positive integer is always greater than that positive integer.)

## Chapter V

# THE CALCULI OF $\lambda$-$K$-CONVERSION AND $\lambda$-$\delta$-CONVERSION

17. THE CALCULUS OF $\lambda$-$K$-CONVERSION. The calculus of $\lambda$-$K$-conversion is obtained if a single change is made in the construction of the calculus of $\lambda$-conversion which appears in §§ 5,6: namely, in the definition of well-formed formula (§5) deleting the words "and contains at least one free occurrence of **x**" from the rule 3. The rules of conversion, I, II, III, in §6 remain unchanged, except that well-formed is understood in the new sense.

Typical of the difference between the calculi of $\lambda$-conversion and $\lambda$-$K$-conversion is the possibility of defining in the latter the constancy function,

$$K \rightarrow \lambda a(\lambda b a),$$

and the integer zero, by analogy with definitions of the positive integers in §8,

$$0 \rightarrow \lambda a(\lambda b b).$$

Many of the theorems of §7 hold also in the calculus of $\lambda$-$K$-conversion. But obvious minor modifications must be made in 7 III and 7 V, and the following theorems fail: 7 XVII, clause (3) of 7 XXVI, and 7 XXXI, and 7 XXXII. Instead of 7 XXXI, the following weaker theorem can be proved, which is sufficient for certain purposes, in particular for the definition of $\mathfrak{p}$ (see §10):

17 I. Let a reduction be called of order one if the application of Rule II involved is a contraction of the initial $(\lambda \mathbf{x} \mathbf{M}) \mathbf{N}_1$ in a formula of the form

$$(\lambda \mathbf{x} \mathbf{M}) \mathbf{N}_1 \mathbf{N}_2 \ldots \mathbf{N}_r \qquad (r = 1, 2, \ldots).$$

Then if $\boldsymbol{A}$ has a normal form, there is a number m such that at most m reductions of order one can occur in a sequence of reductions on $\boldsymbol{A}$.

A notion of <u>λ-K-definability</u> of functions of non-negative integers may be introduced, analogous to that of λ-definability of functions of positive integers, and the developments of Chapter III may then be completely paralleled in the calculus of λ-K-conversion. The same definitions may be employed for the successor function and for addition and multiplication as in Chapter III. Many of the developments are simplified by the presence of the zero: in particular, ordered pairs may be employed instead of ordered triads in the definition of the predecessor function, and the definition of $\mathfrak{p}$ may be simplified as in Turing [58].

It can be proved (see Kleene [37], Turing [57]) that a function $F$ of one non-negative integer argument is λ-K-definable if and only if $\lambda x \,.\, F(x-1)+1$ is λ-definable -- and similarly for functions of more than one argument.

The calculus of λ-K-conversion has obvious advantages over the calculus of λ-conversion, including the possibility of defining the constancy function and of introducing the integer zero in a simpler and more natural way. However, for many purposes -- in particular for the development of a system of symbolic logic such as that sketched in §21 below -- these advantages are more than offset by the failure of 7 XXXII. Indeed if we regard those and only those formulas as meaningful which have a normal form, it becomes clearly unreasonable that $\boldsymbol{FN}$ should have a normal form and $\boldsymbol{N}$ have no normal form (as may happen in the calculus of λ-K-conversion); or even if we impose a more stringent condition of meaningfulness, Rule III of the calculus of λ-K-conversion can be objected to on the ground that if $\boldsymbol{M}$ is a meaningful formula containing no free variables, the substitution of $(\lambda x\boldsymbol{M})\boldsymbol{N}$ for $\boldsymbol{M}$ ought not to be possible unless $\boldsymbol{N}$ is meaningful. This way of putting the matter involves the meanings of the formulas, and thus an appeal to intuition, but corresponding difficulties do appear in the formal developments in certain directions.

§18. THE CALCULUS OF RESTRICTED λ-*K*-CONVERSION. In order to avoid the difficulty just described, Bernays [4] has proposed a modification of the calculus of λ-*K*-conversion which consists in adding to Rules II and III the proviso that **N** shall be in normal form (notice that the condition of being in normal form is effective, although that of having a normal form is not). We shall call the calculus so obtained the calculus of restricted λ-*K*-conversion. In it, as follows by the methods of §7, a formula which in the calculus of λ-*K*-conversion had a normal form and had no parts without normal form will continue to have the same normal form; in particular, no possibility of conversion into a normal form is lost which existed in the calculus of λ-conversion. On the other hand, all of the theorems 7 XXVIII - 7 XXXII remain valid in the calculus of restricted λ-*K*-conversion -- and are much more simply proved than in the calculus of λ-conversion. (It should be added that the content of the theorems 7 XXVIII - 7XXXXII for the calculus of restricted λ-*K*-conversion is in a certain sense much less than the content of these theorems for the calculus of λ-conversion, and in fact cannot be regarded as sufficient to establish the satisfactoriness of the calculus of restricted λ-*K*-conversion from an intuitive viewpoint without addition of such a theorem as that asserting the equivalence to the calculus of (unrestricted) λ-*K*-conversion in the case of formulas all of whose parts have normal forms.)

The development of the calculus of restricted λ-*K*-conversion may follow closely that of the calculus of λ-conversion (as in Chapters II-IV), with such modifications as are indicated in §17 for the calculus of λ-*K*-conversion. Many of the theorems must have added hypotheses asserting that certain of the formulas involved have normal forms.

§19. TRANSFINITE ORDINALS. Church and Kleene [15] have extended the concept of λ-definability to ordinal numbers of the second number class and functions of such ordinal numbers. There results from this on the one hand an extension of the notion of effective calculability to the second number class (cf. Church [13], Kleene [39], Turing [59]), and on the other hand a method of introducing some theory of ordinal numbers into the system of symbolic logic of §21 below.

Instead of reproducing here this development within the calculus of λ-conversion, we sketch briefly an analogous development within the calculus of restricted λ-*K*-conversion.

According to the idea underlying the definitions of §8, the positive integers (or the non-negative integers) are certain functions of functions, namely the finite powers of a function in the sense of iteration. This idea might be extended to the ordinal numbers of the second number class by allowing them to correspond in the same way to the transfinite powers of a function, provided that we first fixed upon a limiting process relative to which the transfinite powers should be taken. Thus the ordinal $\omega$ could be taken as the function whose value for a function $f$ as argument is the function $g$ such that $gx$ is the limit of the sequence, $x, fx, f(fx), \ldots$ . Then $\omega+1$ would be $\lambda x.f(\omega f x)$, and so on.

Or, instead of fixing upon a limiting process, we may introduce the limiting process as an additional argument $a$ (for instance taking the ordinal $\omega$ to be the function whose value for $a$ and $f$ as arguments is the function $g$ such that $gx$ is the limit of the sequence $x, fx, f(fx), \ldots$ , relative to the limiting process $a$). This leads to the following definitions in the calculus of restricted λ-*K*-conversion, the subscript o being used to distinguish these notations from similar notations used in other connections:

$$0_o \rightarrow \lambda a(\lambda b(\lambda c c)),$$

$$1_o \rightarrow \lambda abc.bc,$$

$$2_o \rightarrow \lambda abc.b(bc), \quad \text{and so on.}$$

$$S_o \rightarrow \lambda dabc.b(dabc).$$

$$L_o \rightarrow \lambda rabc.a(\lambda d.rdabc).$$

$$\omega_o \rightarrow \lambda abc.a(\lambda d.dabc).$$

We prescribe that $0_o$ shall represent the ordinal 0; if $\mathbf{N}$ represents the ordinal $n$, the principal normal form of $S_o\mathbf{N}$ shall represent the ordinal $n+1$; if $\mathbf{R}$ represents the monotone increasing infinite sequence of ordinals, $n_0, n_1, n_2, \ldots$, in the sense that $\mathbf{R}0_o, \mathbf{R}1_o, \mathbf{R}2_o, \ldots$ are convertible into formulas representing $n_0, n_1, n_2, \ldots$, respectively, then the principal normal form of $L_0 \mathbf{R}$ shall represent the upper limit of this infinite sequence of ordinals. The transfinite ordinals which are represented by formulas then turn out to constitute a certain segment of the second number class, which may be described as consisting of those ordinals which can be effectively built up to from below (in a sense which we do not make explicit here).

The formula representing a given ordinal of the second number class is not unique: for example, the ordinal $\omega$ is represented not only by $\omega_0$ but also by the principal normal form of $L_0 S_0$, and by many other formulas. Hence the formulas representing ordinals are not to be taken as denoting ordinals but rather as denoting certain things which are in many-one correspondence with ordinals.

A function $F$ of ordinal numbers is said to be <u>$\lambda$-$K$-defined</u> by a formula $\mathbf{F}$ if (1) whenever $Fm = n$ and $\mathbf{M}$ represents $m$, the formula $\mathbf{FM}$ is convertible into a formula representing $n$, and (2) whenever an ordinal $m$ is not in the range of $F$ and $\mathbf{M}$ represents $m$, the formula $\mathbf{FM}$ has no normal form.

The foregoing account presupposes the classical second number class. By suitable modifications (cf. Church [13]), this presupposition may be eliminated, with the result that the calculus of restricted $\lambda$-$K$-conversion is used to obtain a definition of a (non-classical) <u>constructive</u> second number class, in which each classical ordinal is represented, if at all, by an infinity of elements.

20. THE CALCULUS OF $\lambda$-$\delta$-CONVERSION. <u>The calculus of $\lambda$-$\delta$-conversion</u> is obtained by making the following changes in the construction of the calculus of $\lambda$-conversion which appears in §§5, 6: adding to the list of primitive symbols a symbol $\delta$, which is neither an improper symbol nor a variable, but is classed with the variables as a <u>proper symbol</u>; adding to the rule 1 in the definition of <u>well-formed formula</u> that the symbol $\delta$ is a well-formed formula; and adding to the rules of conversion in §6 four additional rules, as follows:

IV. To replace any part $\delta \mathbf{MN}$ of a formula by 1, provided that $\mathbf{M}$ and $\mathbf{N}$ are in $\delta$-normal form and contain no free variables and $M$ is not convertible-I into $N$.

V. To replace any part 1 of a formula by $\delta MN$, provided that $M$ and $N$ are in δ-normal form and contain no free variables and $M$ is not convertible-I into $N$.

VI. To replace any part $\delta MM$ of a formula by 2, provided that $M$ is in δ-normal form and contains no free variables.

VII. To replace any part 2 of a formula by $\delta MM$, provided that $M$ is in δ-normal form and contains no free variables.

Here a formula is said to be in δ-normal form if it contains no part of the form $(\lambda x P)Q$ and contains no part of the form $\delta RS$ with $R$ and $S$ containing no free variables. It is necessary to observe that both the condition of being in δ-normal form and the condition that $M$ is not convertible-I into $N$ are effective.

A conversion (or a λ-δ-conversion) is a finite sequence of applications of Rules I-VII. A λ-δ-conversion is called a reduction (or a λ-δ-reduction) if it contains no application of Rules III, V, VII and exactly one application of one of the Rules II, IV, VI. $A$ is said to be immediately reducible to $B$ if there is a reduction of $A$ into $B$, and $A$ is said to be reducible to $B$ if there is a conversion of $A$ into $B$ which consists of one or more successive reductions.

All the theorems of §7 hold also in the calculus of λ-δ-conversion, if some appropriate modifications are made (see Church and Rosser [16]). The residuals of $(\lambda x_p M_p)N_p$ after an application of Rule I or II are defined in the same way as before, and after an application of IV or VI they are defined as what $(\lambda x_p M_p)N_p$ becomes (this is always something of the form $(\lambda x_p M'_p)N'_p$). The residuals of $\delta M_p N_p$ after an application of I, II, IV, or VI are defined only in the case that $M_p$ and $N_p$ are in δ-normal form and contain no free variables. In that case the residuals of $\delta M_p N_p$ are whatever part or parts of the entire resulting formula $\delta M_p N_p$ becomes, except that after an application of IV or VI in which $\delta M_p N_p$ itself is contracted (i.e., replaced by 1 or 2), $\delta M_p N_p$ has no residual. Thus residuals of $\delta M_p N_p$

are always of the form $\delta MN$, where $M$ and $N$ are in δ-normal form and contain no free variables. A sequence of contractions on a set of parts $(\lambda x_j M_j)N_j$ and $\delta R_1 S_1$ of $A_1$, where $R_1$ and $S_1$ are in δ-normal form and contain no free variables, is defined by analogy with the definition in §7. Similarly a terminating sequence of such contractions. In 7 XXV, the set of parts of $A$ on which a sequence of contractions is taken is allowed to include not only parts of the form $(\lambda x_j M_j)N_j$, but also parts of the form $\delta R_1 S_1$ in which $R_1$ and $S_1$ are in δ-normal form and contain no free variables. The modified 7 XXV may then be proved by an obvious extension of the proof given in §7, and thereupon 7 XXVI - 7 XXXII follow as before. In 7 XXVI - 7 XXXII "conv-I-II" must be replaced throughout by "conv-I-II-IV-VI" and in 7 XXVI the case must also be considered that $A$ imr $B$ by a contraction of the part $\delta MN$ of $A$. For 7 XXX, there must be supplied a definition of principal δ-normal form of a formula, analogous to the definition in §6 of the principal (λ-)normal form.

In connection with the calculus of λ-δ-conversion we shall use both of the terms λ-conversion and λ-δ-conversion, the former meaning a finite sequence of applications of Rules I-III, the latter a finite sequence of applications of Rules I-VII. The term conversion will be used to mean a λ-δ-conversion, as already explained.

Similarly we shall use both of the terms λ-normal form of a formula and δ-normal form of a formula. A formula will be called a λ-normal form of another if it is in λ-normal form and can be obtained from the other by λ-conversion. A formula will be called a δ-normal form of another if it is in δ-normal form and can be obtained from the other by λ-δ-conversion. By 7 XXIX applied to the calculus of λ-conversion, the λ-normal form of a formula (in the calculus of λ-δ-conversion), if it exists, is unique to within applications of Rule I. By the analogue of 7 XXIX for the calculus of λ-δ-conversion, the δ-normal form of a formula, if it exists, is unique to within applications of Rule I.

In order to see that the calculus of λ-δ-conversion requires an intensional interpretation (cf. §2), it is sufficient to observe that, for example, although 1 and $\lambda ab.\delta ab1ab$ correspond to the same function in extension, they are nevertheless not interchangeable, since $\delta 11$ conv 2 but $\delta 1(\lambda ab.\delta ab1ab)$ conv 1.

A constancy function $\varkappa$ may be defined:

$$\varkappa \rightarrow \lambda ab.\delta bbIa.$$

Then $\varkappa AB$ conv $A$, if $B$ has a δ-normal form and contains no free variables, and in that case only (the conversion properties of $\varkappa$ are thus weaker than those of the formula $K$ in either of the calculi of λ-$K$-conversion).

The entire theory of λ-definability of functions of positive integers carries over into the calculus of λ-δ-conversion, since the calculus of λ-conversion is contained in that of λ-δ-conversion as a part. It only requires proof that the notion of λ-δ-definability of functions of positive integers is not more general than that of λ-definability, and this can be supplied by known methods (e.g., those of Kleene [37]).

The theory of combinations carries over into the calculus of λ-δ-conversion, provided that we redefine a combination to mean an {I, J, δ}-combination. In defining the combination belonging to a formula, it is necessary to add the provision that the combination belonging to δ is δ.

If $A_1$ is a well-formed formula of the calculus of λ-δ-conversion and contains no free variables, a formula $B_1$ can be found such that $B_1J$ conv $A_1$ and $B_1I$ conv $I$. For let $A_1'$ be the combination belonging to $A_1$, unless that combination fails to contain an occurrence of either $J$ or δ, in which case let $A_1'$ be $JIIII$. Let $A_1''$ be obtained from $A_1'$ be replacing $J$ and δ throughout by $j$ and $\delta Ij(\lambda x.x(\lambda y.yIII))(\lambda z.zI)\delta$ respectively. Then $B_1$ may be taken as $\lambda jA_1''$.

Hence 14 I, and the remaining theorems of §14, may be proved for the calculus of λ-δ-conversion in the same way as for the calculus of λ-conversion.

In order to obtain a combinatory equivalent of λ-δ-conversion, analogous to the combinatory equivalent of λ-conversion given in §15, it is necessary to add to OI-OXXXVIII the following four additional operations -- where $F$, $A$, $B$, $C$ are combinations, and $A$ and $B$ belong to formulas in δ-normal form, contain no free variables, and are not the same, and $C$ belongs to the formula which represents the Gödel number of $A$:

OXXXIX. $\boldsymbol{F}(\delta \boldsymbol{A}\boldsymbol{B}) \vdash \boldsymbol{F}(\beta I)$.

OXL. $\boldsymbol{F}\boldsymbol{A}\boldsymbol{B}(\beta I) \vdash \boldsymbol{F}\boldsymbol{A}\boldsymbol{B}(\delta \boldsymbol{A}\boldsymbol{B})$.

OXLI. $\boldsymbol{F}(\delta \boldsymbol{A}\boldsymbol{A}) \vdash \boldsymbol{F}(\omega\beta)$.

OXLII. $\boldsymbol{F}\boldsymbol{C}(\omega\beta) \vdash \boldsymbol{F}\boldsymbol{C}(\delta \boldsymbol{A}\boldsymbol{A})$.

The reader should verify that the conditions on $\boldsymbol{A}$, $\boldsymbol{B}$, $\boldsymbol{C}$ -- although complex in character -- are effective (§6).

In order to see that these four operations are equivalent, in the presence of OI - OXXXVIII, to the rules of conversion IV - VII, it is necessary to observe that $\beta I$ and $\omega\beta$ are λ-convertible into 1 and 2 respectively.

To show that OXLII provides an equivalent to Rule VII, we must show that it enables us to change $\boldsymbol{G}(\omega\beta)$ into $\boldsymbol{G}(\delta \boldsymbol{A}\boldsymbol{A})$. Since OI - OXXXVIII are equivalent to λ-conversion, this can be done as follows: $\boldsymbol{G}(\omega\beta)$ is λ-convertible into $\gamma(\tau I)\boldsymbol{G}\boldsymbol{C}(\omega\beta)$, and this becomes, by OXLII, $\gamma(\tau I)\boldsymbol{G}\boldsymbol{C}(\delta \boldsymbol{A}\boldsymbol{A})$, and this in turn is λ-convertible into $\boldsymbol{G}(\delta \boldsymbol{A}\boldsymbol{A})$.

Similarly, to show that OXL provides an equivalent of Rule V, we must show that it enables us to change $\boldsymbol{G}(\beta I)$ into $\boldsymbol{G}(\delta \boldsymbol{A}\boldsymbol{B})$. This can be done as follows: $\boldsymbol{G}(\beta I)$ is λ-convertible into $\gamma(\gamma(\tau I)\boldsymbol{G})(\beta I)(\omega\beta)$; and this can be changed by the method of the preceding paragraph into $\gamma(\gamma(\tau I)\boldsymbol{G})(\beta I)(\delta \boldsymbol{B}\boldsymbol{B})$; and this is λ-convertible into $\gamma(\gamma(\tau I)(\gamma(\gamma(\tau I)\boldsymbol{G})(\beta I)))(\delta \boldsymbol{B}\boldsymbol{B})(\omega\beta)$; and this can be changed by the method of the preceding paragraph into $\gamma(\gamma(\tau I)(\gamma(\gamma(\tau I)\boldsymbol{G})(\beta I)))(\delta \boldsymbol{B}\boldsymbol{B})(\delta \boldsymbol{A}\boldsymbol{A})$; and this is λ-convertible into $\gamma(\beta(\gamma(\beta\gamma(\gamma(\gamma(\beta(\beta\beta)(\omega\delta))I)(\gamma(\gamma(\tau I)\boldsymbol{G})))))(\omega\delta))\boldsymbol{A}\boldsymbol{B}(\beta I)$; and this becomes, by OXL, $\gamma(\beta(\gamma(\beta\gamma(\gamma(\gamma(\beta(\beta\beta)(\omega\delta))I)(\gamma(\gamma(\tau I)\boldsymbol{G})))))(\omega\delta))\boldsymbol{A}\boldsymbol{B}(\delta \boldsymbol{A}\boldsymbol{B})$; and this is λ-convertible into $\gamma(\gamma(\tau I)(\gamma(\gamma(\tau I)\boldsymbol{G})(\delta \boldsymbol{A}\boldsymbol{B})))(\delta \boldsymbol{B}\boldsymbol{B})(\delta \boldsymbol{A}\boldsymbol{A})$; and this becomes, by OXLI, $\gamma(\gamma(\tau I)(\gamma(\gamma(\tau I)\boldsymbol{G})(\delta \boldsymbol{A}\boldsymbol{B})))(\delta \boldsymbol{B}\boldsymbol{B})(\omega\beta)$; and this is λ-convertible into $\gamma(\gamma(\tau I)\boldsymbol{G})(\delta \boldsymbol{A}\boldsymbol{B})(\delta \boldsymbol{B}\boldsymbol{B})$; and this becomes, by OXLI, $\gamma(\gamma(\tau I)\boldsymbol{G})(\delta \boldsymbol{A}\boldsymbol{B})(\omega\beta)$; and this, finally, is λ-convertible into $\boldsymbol{G}(\delta \boldsymbol{A}\boldsymbol{B})$.

Only minor modifications are necessary in §16 in order to carry over its results to the calculus of λ-δ-conversion. In the definition of the Gödel number of a combination the clause must be added: (2a) The Gödel number of δ is 5. In the construction of the formula, form, it is only necessary to impose on $\mathfrak{D}$ the further condition that $\mathfrak{D}$ 5 conv $\lambda x.x12\delta$, so insuring that form 5 conv δ. The construction of o remains unchanged. The formulas occ, ε, $\mathfrak{E}$, r, Λ, prim, norm, and $0_1$ - $0_{38}$ may then be obtained, having the properties described in §16 (norm λ-defines the property of a combination of belonging to a formula which is in λ-normal form). The formulas cb, ncb, 0, cnvt, nf (the λ-normal form of), and ν may then also be obtained as before. The formula, cb, represents an enumeration of the least set of formulas which contains 1, 3, and 5 and is closed under the operation of forming $(\lambda ab\,.\,2 \times \text{nr}\,ab)XY$ from the formulas $X$, $Y$.

Besides norm it is also possible to obtain a formula, dnorm, which λ-defines the property of a combination of belonging to a formula in δ-normal form. Details of this are left to the reader.

Formulas $0_{39}$- $0_{42}$ may be obtained, related to the operations OXXXIX - OXLII in the same way that $0_1$ - $0_{38}$ are related to OI - OXXXVIII. We give details in the case of $0_{40}$ and $0_{42}$. Let $\mathfrak{F}_{40}$ be a formula such that $\mathfrak{F}_{40}1$ conv $I$ and $\mathfrak{F}_{40}2$ conv $\lambda x\,.\,2 \times \text{nr}\; x_1[2 \times \text{nr}[2 \times \text{nr}\; 5x_{112}]x_{12}]$; then let

$$0_{40} \rightarrow \lambda x\,.\,\mathfrak{F}_{40}[\text{par}\; x + \text{par}\; x_1 + \text{par}\; x_{11} + \text{prim}\; x_{112} + \text{dnorm}\; x_{112} + \text{prim}\; x_{12} + \text{dnorm}\; x_{12} + \text{eq}\; \eta x_2 \dot{-} \text{eq}\; x_{112}x_{12} \dot{-} \overline{13}]x,$$

$\eta$ being the formula representing the Gödel number of $\beta I$. Let $\mathfrak{F}_{42}$ be a formula such that $\mathfrak{F}_{42}1$ conv $I$ and $\mathfrak{F}_{42}2$ conv $\lambda x\,.\,2 \times \text{nr}\; x_1[2 \times \text{nr}\;[2 \times \text{nr}\; 5(\text{form}\; x_{12})](\text{form}\; x_{12})]$; then let

$$0_{42} \rightarrow \lambda x\,.\,\mathfrak{F}_{42}[\text{par}\; x + \text{par}\; x_1 + h(\nu x_{12})x_{12} + \text{eq}\; \zeta x_2 \dot{-} 6]x,$$

where $\zeta$ is the formula representing the Gödel number of $\omega\beta$, and $h$ is such a formula that $h1$ conv $\lambda x.x1$ and $h2$ conv $\lambda x.\text{min}\;(\text{prim}\;(\text{form}\; x))(\text{dnorm}\;(\text{form}\; x))$.

Then a formula, do, may be obtained, analogous to 0 but involving all of $0_1$ - $0_{42}$ instead of only $0_1$ - $0_{38}$. Let

$$\text{dcnvt} \rightarrow \lambda ab\,.\,\text{do}\; ba.$$

Then, if $C$ represents the Gödel number of a combination $M$, the formula, dcnvt $C$, enumerates the Gödel numbers of combinations obtainable from $M$ by λ-δ-conversion (whereas cnvt $C$

enumerates merely the Gödel numbers of combinations obtainable from $\boldsymbol{M}$ by λ-conversion).

It is also possible, by using the formula, dnorm, to obtain a formula, dnf, which λ-defines the operation δ-normal form of a formula, and a formula, dncb, which enumerates the Gödel numbers of combinations which belong to formulas in δ-normal form and contain no free variables. The definitions parallel those of nf and ncb.

Finally, in the calculus of λ-δ-conversion, a formula, met, may be obtained which provides a kind of inverse of the function, form: if $\boldsymbol{M}$ is a formula which contains no free variables and has a δ-normal form, then met $\boldsymbol{M}$ is convertible into the formula representing the Gödel number belonging to the δ-normal form of $\boldsymbol{M}$. The definition is as follows:

$$\text{met} \rightarrow \lambda x\ .\ \text{dncb}\ (\mathfrak{p}(\lambda n\ .\ \delta(\text{form}\ (\text{dncb}\ n))x)1).$$

21. A SYSTEM OF SYMBOLIC LOGIC. If we identify the truth values, truth and falsehood, with the positive integers 2 and 1 respectively, we may base a system of symbolic logic on the calculus of λ-δ-conversion. This system has one primitive formula or axiom, namely the formula 2, and seven rules of inference, namely the rules I - VII of λ-δ-conversion; the provable formulas, or theses, of the system are the formulas which can be derived from the formula 2 by sequences of applications of the rules of inference. (As a matter of fact, the rules of inference II, IV, VI are superfluous, in the sense that their omission would not decrease the class of provable formulas, as follows from 7 XXVII, or rather from the analogue of this theorem for the calculus of λ-δ-conversion.)

The identification of the truth values, truth and falsehood, with the positive integers 2 and 1 is, of course, artificial, but apparently it gives rise to no actual formal difficulty. If it be thought objectionable, the artificiality may be avoided by a minor modification in the system, which consists in introducing a symbol ⊢ and writing ⊢2, instead of 2, as the primitive formula; all the theses of the system will then be preceded by the sign ⊢, which may be interpreted as asserting that that which follows is equal to 2.

In this system of symbolic logic the fundamental operations of the propositional calculus -- negation, conjunction, disjunction -- may be introduced by the following definitions:

$$[\sim A] \longrightarrow \pi(\lambda \mathfrak{a}.\mathfrak{a} I(\delta 2 A))(\lambda \mathfrak{a}.\mathfrak{a} I(\delta 1 A)).$$

$$[A \& B] \longrightarrow 4 \dot{-} . [\sim A] + [\sim B].$$

$$[A \vee B] \longrightarrow \sim . [\sim A] \& [\sim B].$$

It follows from these definitions that $A \vee B$ cannot be a thesis unless either $A$ or $B$ is a thesis -- and this situation apparently cannot be altered by any suitable change in the definitions. Since this property is known to fail for classical systems of logic, e.g., that of Whitehead and Russell's Principia Mathematica, it is clear that the present system therefore differs from the classical systems in a direction which may be regarded as finitistic in character.

Functions of positive integers are of course represented in the system by the formulas λ-defining these functions, and properties of and relations between positive integers are represented by the formulas λ-defining the corresponding characteristic functions. The propositional function to be a positive integer is represented in the system as a formula $N$, defined as follows (referring to §§16, 20):

$$N \longrightarrow \lambda x.\nu(\text{met } x).$$

The general relation of equality or identity (in intension) is represented by $\delta$.

An existential quantifier $\Sigma$ may be introduced:

$$\iota \longrightarrow \lambda f . \text{form } (Z'(H(\text{dcnvt } \alpha(\mathfrak{p}(\lambda n . \delta f$$
$$(\text{form } (Z(H(\text{dcnvt } \alpha n)))))1)))),$$

where $\alpha$ is the formula representing the Gödel number belonging to the formula 2;

$$\Sigma \longrightarrow \lambda f.f(\iota f).$$

Here $\iota$ represents a general selection operator. Given a formula $F$; if there is any formula $A$ such that $FA$ conv 2, then $\iota F$ is one of the formulas $A$ having this property; and in the contrary case $\iota F$ has no normal form. Consequently $\Sigma$ represents an existential quantifier without a negation: $\Sigma F$ conv 2 if there is a formula $A$ such that $FA$ conv 2, and in the contrary case $\Sigma F$ has no normal form.

The operator $\iota$ should be compared with Hilbert's operator $\epsilon$ [31 and elsewhere], or, perhaps better, the $\eta$-operator of Hilbert and Bernays [33]. The $\iota$ should be used with the caution that the equivalence of propositional functions represented in the system by $F$ and $G$ need not imply the equality of $\iota F$ and $\iota G$.

The interpretation of $\iota$ as a selection operator and of $\Sigma$ as an existential quantifier depends on an identification of formal provability in the system with truth. But this is justified by a completeness property which the system possesses: a formula which is not provable, unless it is convertible into a principal normal form other than 2 and hence is disprovable, must have no normal form, and hence be meaningless.

For convenience in the further development of the system, or for the sake of comparison with more usual notations, we may introduce the abbreviations:

$$[\iota x M] \rightarrow \iota(\lambda x M).$$

$$[\exists x M] \rightarrow \Sigma(\lambda x M).$$

The problem of introducing universal quantifiers into the system, or, equivalently, of introducing existential quantifiers having a negation, is beyond the scope of the present treatise. It follows by the methods of Gödel [27] that any universal quantifier introduced by definition will have a certain character of incompleteness; this is in effect the same incompleteness property which, in accordance with the results of Gödel, almost any consistent and satisfactorily adequate system of formal logic must have, except that it here appears transferred from the realm of provability to the realm of meaning of the quantifiers.

The consistency of the system of symbolic logic just outlined is a corollary of 7 XXX, or rather of the analogue of this theorem for the calculus of λ-δ-conversion. This consistency proof is of a strictly constructive or finitary nature.

(The failure in this system of the known paradoxes of set theory depends, in some of the simpler cases, merely on the fact that the formula which would otherwise lead to the paradox fails to have a normal form. Thus, in the case of Russell's paradox, we find that $(\lambda x.\sim(xx))(\lambda x.\sim(xx))$ has no normal form; and in the case of Grelling's paradox concerning heterological words, or, as we shall put it, concerning heterological Gödel numbers, we find that $(\lambda x.\sim(\text{form } xx))(\text{met } (\lambda x.\sim(\text{form } xx)))$ has no normal form. In more complicated cases, where the expression of the paradox requires a universal quantifier, the failure may depend on the above indicated incompleteness property of the quantifier.)

## INDEX OF THE PRINCIPAL FORMULAS INTRODUCED BY DEFINITION

§5. $I$, $[M+N]$, $[M \times N]$, $[M^N]$.

§8. 1, 2, 3, ..., $S$, $[M+N]$, $[M \times N]$, $[M^N]$.

§9. $[M, N]$, $[L, M, N]$, $2_1$, $2_2$, $3_1$, $3_2$, $3_3$, $P$, $[M \dot{-} N]$, min, max, par, $H$, $Z$, $Z'$, nr.

§10. exc, eq, $\mathfrak{p}$, $\mathfrak{P}$, $\pi$.

§12. $I$, $J$, $\tau$.

§13. $B$, $C$, $W$, $T$, $D$.

§15. $\gamma$, $\beta$, $\omega$.

§16. form, o, occ, $\mathfrak{e}$, $\mathfrak{C}$, $\mathfrak{r}$, $\Lambda$, prim, norm, $0_1$ - $0_{38}$, cb, ncb, 0, cnvt, nf, $\nu$.

§17. $K$, o.

§19. $0_0$, $1_0$, $2_0$, ..., $S_0$, $L_0$, $\omega_0$.

§20. $\kappa$, form, o, occ, $\mathfrak{e}$, $\mathfrak{C}$, $\mathfrak{r}$, $\Lambda$, prim, norm, $0_1$ - $0_{38}$, cb, ncb, 0, cnvt, nf, $\nu$, dnorm, $0_{39}$ - $0_{42}$, do, dcnvt, dnf, dncb, met.

§21. $[\sim A]$, $[A \& B]$, $[A \vee B]$, $N$, $\iota$, $\Sigma$, $[\iota x M]$, $[\exists x M]$.

BIBLIOGRAPHY


1. Wilhelm Ackermann, Zum Hilbertschen Aufbau der reellen Zahlen, Mathematische Annalen, vol. 99 (1928), pp. 118 - 133.

2. Paul Bernays, Sur le platonisme dans les mathématiques, l'Enseignement mathématique, vol. 34 (1935), pp. 52 - 69.

3. Paul Bernays, Quelques points essentiels de la métamathématique, ibid., pp. 70 - 95.

4. Paul Bernays, Review of Church and Rosser [16], The journal of symbolic logic, vol. 1 (1936), pp. 74 - 75.

5. Alonzo Church, A set of postulates for the foundation of logic, Annals of mathematics, ser. 2, vol. 33 (1932), pp. 346 - 366.

6. Alonzo Church, A set of postulates for the foundation of logic (second paper), ibid., ser. 2, vol. 34 (1933), pp. 839 - 864.

7. Alonzo Church, The Richard paradox, the American mathematical monthly, vol. 41 (1934), pp. 356 - 361.

8. Alonzo Church, A proof of freedom from contradiction, Proceedings of the National Academy of Sciences of the United States of America, vol. 21 (1935), pp. 275 - 281.

9. Alonzo Church, An unsolvable problem of elementary number theory, American journal of mathematics, vol. 58, (1936), pp. 345 - 363.

10. Alonzo Church, Mathematical logic, mimeographed lecture notes, Princeton University, 1936.

11. Alonzo Church, A note on the Entscheidungsproblem, The journal of symbolic logic, vol. 1 (1936), pp. 40 - 41.

12. Alonzo Church, Correction to A note on the Entscheidungs problem, ibid., pp. 101 - 102.

13. Alonzo Church, The constructive second number class, Bulletin of the American Mathematical Society, vol. 44 (1938), pp. 224 - 232.

14. Alonzo Church, On the concept of a random sequence, ibid., vol. 46 (1940), pp. 130 - 135.

15. Alonzo Church and S. C. Kleene, Formal definitions in the theory of ordinal numbers, Fundamenta mathematicae, vol. 28 (1937), pp. 11 - 21.

16. Alonzo Church and J. B. Rosser, Some properties of conversion, Transactions of the American Mathematical Society, vol. 39 (1936), pp. 472 - 482.

17. H. B. Curry, An analysis of logical substitution, American journal of mathematics, vol. 51 (1929), pp. 363 - 384.

18. H. B. Curry, Grundlagen der kombinatorischen Logik, ibid., vol. 52 (1930), pp. 509 - 536, 789 - 834.

19. H. B. Curry, The universal quantifier in combinatory logic, Annals of mathematics, ser. 2, vol. 32 (1931), pp. 154 - 180.

20. H. B. Curry, Some additions to the theory of combinators, American journal of mathematics, vol. 54 (1932), pp. 551 - 558.

21. H. B. Curry, Apparent variables from the standpoint of combinatory logic, Annals of mathematics, ser. 2, vol. 34 (1933), pp. 381 - 404.

22. H. B. Curry, Some properties of equality and implication in combinatory logic, ibid., ser. 2, vol. 35 (1934), pp. 849 - 860.

23. H. B. Curry, Functionality in combinatory logic, Proceedings of the National Academy of Sciences of the United States of America, vol. 20 (1934), pp. 584 - 590.

24. H. B. Curry, First properties of functionality in combinatory logic, The Tôhoku mathematical journal, vol. 41 (1936), pp. 371 - 401.

25. H. B. Curry, Review of Church [10], The journal of symbolic logic, vol. 2 (1937), pp. 39 - 40.

26. Frederic B. Fitch, A system of formal logic without an analogue to the Curry $W$ operator, The journal of symbolic logic, vol. 1 (1936), pp. 92 - 100.

27. Kurt Gödel, Über formal unentscheidbare Sätze der Principia Mathematica und verwandter Systeme I, Monatshefte für Mathematik und Physik, vol. 38 (1931), pp. 173 - 198.

28. Kurt Gödel, On undecidable propositions of formal mathematical systems, mimeographed lecture notes, The Institute for Advanced Study, 1934.

29. Kurt Gödel, Über die Länge von Beweisen, Ergebnisse eines mathematischen Kolloquiums, no. 7 (1936), pp. 23 - 24.

30. Jaques Herbrand, Sur la non-contradiction de l'arithmétique, Journal fur die reine und angewandte Mathematik, vol. 166 (1931), pp. 1 - 8.

31. David Hilbert, Über das Unendliche, Mathematische Annalen, vol. 95 (1926), pp. 161 - 190.

32. David Hilbert and Paul Bernays, Grundlagen der Mathematik, vol. 1, Julius Springer, Berlin, 1934.

33. David Hilbert and Paul Bernays, Grundlagen der Mathematik, vol. 2, Julius Springer, Berlin, 1939.

34. S. C. Kleene, Proof by cases in formal logic, Annals of mathematics, ser. 2, vol. 35 (1934), pp. 529 - 544.

35. S. C. Kleene, A theory of positive integers in formal logic, American journal of mathematics, vol. 57 (1935), pp. 153 - 173, 219 - 244.

36. S. C. Kleene, General recursive functions of natural numbers, Mathematische Annalen, vol. 112 (1936), pp. 727 - 742; see [45], and [39] footnote 4.

37. S. C. Kleene, $\lambda$-definability and recursiveness, Duke mathematical journal, vol. 2 (1936), pp. 340 - 353.

38. S. C. Kleene, A note on recursive functions, Bulletin of the American Mathematical Society, vol. 42 (1936), pp. 544 - 546.

39. S. C. Kleene, On notation for ordinal numbers, The journal of symbolic logic, vol. 3 (1938), pp. 150 - 155.

40. S. C. Kleene and J. B. Rosser, The inconsistency of certain formal logics, Annals of mathematics, ser. 2, vol. 36 (1935) pp. 630 - 636.

41. Rózsa Péter, Über den Zusammenhang der verschiedenen Begriffe der rekursiven Funktion, Mathematische Annalen, vol. 110 (1934), pp. 612 - 632.

42. Rózsa Péter, Konstruktion nichtrekursiver Funktionen, ibid., vol. 111 (1935), pp. 42 - 60.

43. Rózsa Péter, A rekurzív függvények elméletéhez (Zur Theorie der rekursiven Funktionen), Matematikai és fizikai lapok, vol. 42 (1935), pp. 25 - 49.

44. Rózsa Péter, Über die mehrfache Rekursion, Mathematische Annalen, vol. 113 (1936), pp. 489 - 527.

45. Rózsa Péter, Review of Kleene [36], The journal of symbolic logic, vol. 2 (1937), p. 38; see Errata, ibid., vol. 4, (1939), p. iv.

46. Emil L. Post, Finite combinatory processes - formulation 1, ibid., vol. 1 (1936), pp. 103 - 105.

47. J. B. Rosser, A mathematical logic without variables, Annals of mathematics, ser. 2, vol. 36 (1935), pp. 127 - 150, and Duke mathematical journal, vol. 1 (1935), pp. 328 - 355.

48. J. B. Rosser, Extensions of some theorems of Gödel and Church, The journal of symbolic logic, vol. 1 (1936), pp. 87 - 91.

49. Moses Schönfinkel, Über die Bausteine der mathematischen Logik, Mathematische Annalen, vol. 92 (1924), pp. 305 - 316.

50. Thoralf Skolem, Begründung der elementaren Arithmetik durch die rekurrierende Denkweise ohne Anwendung scheinbarer Veränderlichen mit unendlichem Ausdehnungsbereich, Skrifter utgit av Videnskapsselskapet i Kristiania, I. Matematisk-naturvidenskabelig klasse 1923, no. 6.

51. Thoralf Skolem, Über die Zurückführbarkeit einiger durch Rekursionen definierter Relationen auf "arithmetische", Acta scientiarum mathematicarum, vol. 8 (1937) pp. 73 - 88.

52. G. Sudan, Sur le nombre transfini $\omega^\omega$, Bulletin mathématique de la Société Roumaine des Sciences, vol. 30 (1927) pp. 11 - 30.

53. Alfred Tarski, Pojęcie prawdy w językach nauk dedukcyjnych, Travaux de la Société des Sciences et des Lettres de Varsovie, Classe III, Sciences mathématiques et physiques, no. 34, Warsaw 1933.

54. Alfred Tarski, Der Wahrheitsbegriff in den formalisierten Sprachen, German translation of [53] with added Nachwort, Studia philosophica, vol. 1 (1935), pp. 261 - 405.

55. A. M. Turing, On computable numbers, with an application to the Entscheidungsproblem, Proceedings of the London Mathematical Society, ser. 2, vol. 42 (1936), pp. 230-- 265.

56. A. M. Turing, On computable numbers, with an application to the Entscheidungsproblem, A correction, ibid., ser. 2, vol. 43 (1937), pp. 544 - 546.

57. A. M. Turing, Computability and $\lambda$-definability, The journal of symbolic logic, vol. 2 (1937), pp. 153 - 163.

58. A. M. Turing, The $\mathfrak{p}$-function in $\lambda$-$K$-conversion, ibid., p. 164.

59. A. M. Turing, Systems of logic based on ordinals, Proceedings of the London Mathematical Society, ser. 2, vol. 45 (1939), pp. 161 - 228.

Addenda

60. Alonzo Church, A formulation of the simple theory of types, The journal of symbolic logic, vol. 5 (1940), pp. 56 - 68.

61. H. B. Curry, A formalization of recursive arithmetic, American journal of mathematics, vol. 63 (1941), pp.263-282.

62. H. B. Curry, A revision of the fundamental rules of combinatory logic, The journal of symbolic logic, vol. 6 (1941), pp. 41 - 53.

63. H. B. Curry, Consistency and completeness of the theory of combinators, ibid., pp. 54-61.

Further addenda (1951)

64. Alonzo Church, Review of Post [101], The journal of symbolic logic, vol. 8 (1943), pp. 50-52; see erratum, ibid., p. iv. See note thereon by Post, Bulletin of the American Mathematical Society, vol. 52 (1946), p. 264.

65. Paul Csillag, Eine Bemerkung zur Auflösung der eingeschachtelten Rekursion, Acta scientiarum mathematicarum, vol. 11 (1947), pp. 169-173.

66. H. B. Curry, The paradox of Kleene and Rosser, Transactions of the American Mathematical Society, vol. 50 (1941), pp. 454-516.

67. H. B. Curry, The combinatory foundations of mathematical logic, The journal of symbolic logic, vol. 7 (1942), pp. 49-64; see erratum, ibid., vol. 8, p. iv.

68. H. B. Curry, The inconsistency of certain formal logics, ibid., vol. 7 (1942), pp. 115-117; see erratum, ibid., p. iv.

69. H. B. Curry, Some advances in the combinatory theory of quantification, Proceedings of the National Academy of Sciences of the United States of America, vol. 28 (1942), pp. 564-569.

70. H. B. Curry, A simplification of the theory of combinators, Synthese, vol. 7 no. 6A (1949), pp. 391-399.

71. Martin Davis, On the theory of recursive unsolvability, dissertation, Princeton 1950.

72. Robert Feys, La technique de la logique combinatoire, Revue philosophique de Louvain, vol. 44 (1946), pp. 74-103, 237-270.

73. Frederic B. Fitch, A basic logic, The journal of symbolic logic, vol. 7 (1942), pp. 105-114; see erratum, ibid., p. iv.

74. Frederic B. Fitch, Representations of calculi, ibid., vol. 9 (1944), pp. 57-62; see errata, ibid., p. iv.

75. Frederic B. Fitch, A minimum calculus for logic, ibid., pp. 89-94; see erratum, ibid., vol. 10, p. iv.

76. Frederic B. Fitch, An extension of basic logic, ibid., vol. 13 (1948), pp. 95-106.

77. Frederic B. Fitch, <u>The Heine-Borel theorem in extended basic logic</u>, ibid., vol. 14 (1949), pp. 9-15.

78. Frederic B. Fitch, <u>On natural numbers, integers, and rationals</u>, ibid., pp. 81-84.

79. Frederic B. Fitch, <u>A further consistent extension of basic logic</u>, ibid., pp. 209-218.

80. Frederic B. Fitch, <u>A demonstrably consistent mathematics -- Part I</u>, ibid., vol. 15 (1950), pp. 17-24.

81. Hans Hermes, <u>Definite Begriffe und berechenbare Zahlen</u>, Semester-Berichte (Münster i. W.), summer 1937, pp. 110-123.

82. S. C. Kleene, <u>Recursive predicates and quantifiers</u>, Transactions of the American Mathematical Society, vol. 53 (1943), pp. 41-73.

83. S. C. Kleene, <u>On the forms of the predicates in the theory of constructive ordinals</u>, American journal of mathematics, vol. 66 (1944), pp. 41-58.

84. S. C. Kleene, <u>On the interpretation of intuitionistic number theory</u>, The journal of symbolic logic, vol. 10 (1945), pp. 109-124.

85. S. C. Kleene, <u>On the intuitionistic logic</u>, Proceedings of the Tenth International Congress of Philosophy, North-Holland Publishing Company, Amsterdam 1949, pp. 741-743.

86. A. Markoff, <u>On the impossibility of certain algorithms in the theory of associative systems</u>, Comptes rendus (Doklady) de l'Académie des Sciences de l'URSS, n.s. vol. 55 no. 7 (1947), pp. 583-586.

87. A. Markoff, <u>Névozmožnost' nékotoryh algorifmov v téorii associativnyh sistém</u>, Doklady Akadémii Nauk SSSR, vol. 55 (1947), pp. 587-590, vol. 58 (1947), pp. 353-356, and vol. 77 (1951), pp. 19-20.

88. A. Markoff, <u>O nékotoryh nérazréšimyh problémah kasaúščihsá matric</u>, ibid., vol. 57 (1947), pp. 539-542.

89. A. Markoff, <u>O prédstavlénii rékursivnyh funkcij</u>, ibid., vol. 58 (1947), pp. 1891-1892.

90. Andrzej Mostowski, <u>On definable sets of positive integers</u>, Fundamenta mathematicae, vol. 34 (1946), pp. 81-112.

91. Andrzej Mostowski, On a set of integers not definable by means of one-quantifier predicates, Annales de la Société Polonaise de Mathématique, vol. 21 (1948), pp. 114-119.

92. Andrzej Mostowski, Sur l'interprétation géométrique et topologique des notions logiques, Proceedings of the Tenth International Congress of Philosophy, North-Holland Publishing Company, Amsterdam 1949, pp. 767-769.

93. John R. Myhill, Note on an idea of Fitch, The journal of symbolic logic, vol. 14 (1949), pp. 175-176.

94. David Nelson, Recursive functions and intuitionistic number theory, Transactions of the American Mathematical Society, vol. 61 (1947), pp. 307-368; see errata, ibid., p. 556.

95. David Nelson, Constructible falsity, The journal of symbolic logic, vol. 14 (1949), pp. 16-26.

96. M. H. A. Newman, On theories with a combinatorial definition of "equivalence," Annals of mathematics, ser. 2, vol. 43 (1942), pp. 223-243.

97. M. H. A. Newman, Stratified systems of logic, Proceedings of the Cambridge Philosophical Society, vol. 39 (1943), pp. 69-83.

98. Rózsa Péter, Zusammenhang der mehrfachen und transfiniten Rekursionen, The journal of symbolic logic, vol. 15 (1950), pp. 248-272.

99. Rózsa Péter, Zum Begriff der rekursiven reellen Zahl, Acta scientiarum mathematicarum, vol. 12 part A (1950), pp. 239-245.

100. Rózsa Péter, Rekursive Funktionen, Akademischer Verlag, Budapest, 1951.

101. Emil L. Post, Formal reductions of the general combinatorial decision problem, American journal of mathematics, vol. 65 (1943), pp. 197-215.

102. Emil L. Post, Recursively enumerable sets of positive integers and their decision problems, Bulletin of the American Mathematical Society, vol. 50 (1944), pp. 284-316.

103. Emil L. Post, Recursive unsolvability of a problem of Thue, The journal of symbolic logic, vol. 11 (1946), pp. 1-11.

104. Emil L. Post, Note on a conjecture of Skolem, ibid., pp. 73-74.

105. Julia Robinson, Definability and decision problems in arithmetic, The journal of symbolic logic, vol. 14 (1949), pp. 98-114.

106. Julia Robinson, General recursive functions, Proceedings of the American Mathematical Society, vol. 1 (1950), pp. 703-718.

107. Raphael M. Robinson, Primitive recursive functions, Bulletin of the American Mathematical Society, vol. 53 (1947), pp. 925-942.

108. Raphael M. Robinson, Recursion and double recursion, ibid., vol. 54 (1948), pp. 987-993.

109. J. B. Rosser, Review of this monograph, The journal of symbolic logic, vol. 6 (1941), p. 171.

110. J. B. Rosser, New sets of postulates for combinatory logics, ibid., vol. 7 (1942), pp. 18-27; see errata, ibid., vol. 7, p. iv, and vol. 8, p. iv.

111. Thoralf Skolem, Einfacher Beweis der Unmöglichkeit eines allgemeinen Lösungsverfahrens für arithmetische Probleme, Det Kongelige Norske Videnskabers Selskab, Forhandlinger, vol. 13 (1940), pp. 1-4.

112. Thoralf Skolem, Remarks on recursive functions and relations, ibid., vol. 17 (1944), pp. 89-92.

113. Thoralf Skolem, Some remarks on recursive arithmetic, ibid., pp. 103-106.

114. Thoralf Skolem, A note on recursive arithmetic, ibid., pp. 107-109.

115. Thoralf Skolem, Some remarks on the comparison between recursive functions, ibid., pp. 126-129.

116. Thoralf Skolem, Den rekursive aritmetikk, Norsk matematisk tidsskrift, vol. 28 (1946), pp. 1-12.

117. Thoralf Skolem, The development of recursive arithmetic, Den 10. Skandinaviske Matematiker Kongres, Jul. Gjellerups Forlag, Copenhagen 1947, pp. 1-16.

118. Ernst Specker, Nicht konstruktiv beweisbare Sätze der Analysis, The journal of symbolic logic, vol. 14 (1949), pp. 145-158.

119. Alfred Tarski, Andrzej Mostowski and Alfred Tarski, Julia Robinson, abstracts in The journal of symbolic logic, vol. 14 (1949), pp. 75-78.

## CORRECTION AND ADDITIONS

Page 75, line 12. For "Jaques," read "Jacques."

On page 46 the amendment should also be taken into account which is suggested by Rosser [109]. The following simpler expression for $W$ is available:

$$W \mathrm{conv}\ B(T(B(BDB)T))(BBT).$$

Hence replace line 9 on page 46 by this.

In §15, pages 49-51, the combinatory equivalent of conversion which is given can be simplified by the method of Rosser [110], and in particular the proof of the equivalence to conversion can be greatly shortened. Details of this, including the proof of equivalence, may be obtained from Rosser's paper; and the formula 0 of §16, and the formula do of §20, may then be modified correspondingly.

For a combinatory equivalent of $\lambda$-$K$-conversion, and also of $\lambda$-$K$-conversion with the addition of a rule by which $BI$ and $I$ are interchangeable, see [70] -- where Curry employs Rosser's method in order to simplify his earlier treatments of the theory of combinators (which are referred to at the end of §15).

Lightning Source UK Ltd.
Milton Keynes UK
UKHW010711091222
413649UK00001B/56

9 780691 083940